6th International Symposium on Hydrogen Energy and Energy Technologies (HEET 2023)

Journal of Physics: Conference Series Volume 2723

Osaka, Japan
10-11 November 2023

ISBN: 978-1-7138-9278-6
ISSN: 1742-6588

Printed from e-media with permission by:

Curran Associates, Inc.
57 Morehouse Lane
Red Hook, NY 12571

Some format issues inherent in the e-media version may also appear in this print version.

Institute of Physics
Dirac House, Temple Back
Bristol BS1 6BE UK

Phone: 44 1 17 929 7481
Fax: 44 1 17 920 0979

techtracking@iop.org

Additional copies of this publication are available from:

Curran Associates, Inc.
57 Morehouse Lane
Red Hook, NY 12571 USA
Phone: 845-758-0400
Fax: 845-758-2633
Email: curran@proceedings.com
Web: www.proceedings.com

TABLE OF CONTENTS

Journal of Physics: Conference Series **2723** (2024) 011001 doi:10.1088/1742-6596/2723/1/011001

Preface

2023 The 6th International Symposium on Hydrogen Energy and Energy Technologies (HEET 2023)
November 10-11, 2023, Osaka, Japan (virtual conference (ZOOM))

We are glad to introduce you that the 2023 The 6th International Symposium on Hydrogen Energy and Energy Technologies (HEET 2023) was successfully held on November 10-11, 2023.

The vision of the HEET 2023 is a premier interdisciplinary platform for researchers, practitioners and educators to present and discuss the most recent innovations, trends, and concerns as well as practical challenges encountered and solutions adopted in the fields of Hydrogen Energy & Energy Technologies.

During the conference, the conference model was divided into three sessions, including oral presentations, keynote speeches and online Q&A discussion. We have 5 keynote speeches, 1 invited talk and 20 oral presentations both from local and international presenters. The international presenters come from 20 countries and regions namely Russia, Saudi Arabia, Australia, Italy, Slovenia, America, China, Argentina, India, Poland, Nigeria, French and Taiwan. The time spent delivering the talks by each keynote speaker was 30 minutes, the oral presentation speaker was 15 minutes following by a Question Answer (Q & A) Session 5 minutes for each.

We are thankful to all the authors who submitted papers and the delegates for their participation and their interest in HEET as a platform to share their ideas and innovation.

Also, we are also thankful to all the committee members for providing continuous guidance and support and Efforts taken by peer reviewers contributed to improve the quality of papers provided constructive critical comments, improvements and corrections to the authors are gratefully appreciated for their contribution to the success of the conference.

Dr. Jianjun Cai
Editor of the HEET2023 Proceeding
School of Architecture and Traffic Engineering
Guilin University of Electronic Technology
caijianjun@mails.guet.edu.cn https://www.guet.edu.cn/jjxy/info/1479/4165.htm

Journal of Physics: Conference Series **2723** (2024) 011001 doi:10.1088/1742-6596/2723/1/011001

Technical Program Committees

Conference Program Chair

Assoc. Prof. Dr. Yanhui Li, Xi'an JiaoTong University, China

Technical Program Committees

Prof. Dr. Kenshi Itaoka, Kyushu University, Japan

Prof. Dr. S.A. Grigoriev, National Research University "Moscow Power Engineering Institute", Russia

Prof. Dr. Bashir Ahmmad Arima, Yamagata University, Japan

Dr. Jianjun Cai, School of Architecture and Traffic Engineering, Guilin University of Electronic Technology, China

Prof. Dr. Narottam Das, University of Southern Queensland, Australia

Prof. Dr. Rui Chibante, School of Engineering of Polytechnic Institute of Porto, Portugal

Prof. Dr. Balasubramanian Viswanathan, Indian Institute of Technology Madras, India

Prof. Dr. Salvador Aceves, Lawrence Livermore National Laboratory, USA

Prof. Dr. Gheorghe-Daniel Andreescu, Politehnica University of Timisoara, Romania

Prof. Dr. Paolo Beccarelli, University of Nottingham, England

Prof. Dr. Samo Kralj, University of Maribor, Slovenia

Prof. Dr. Yangkyun Kim, Korea Institute of Civil Engineering and Building Technology, Korea

Prof. Dr. Bimal K. Bose, University of Tennessee, USA

Prof. Dr. Jennifer Wen, University of Warwick, UK

Dr. Dai-Viet N. Vo,Universiti Malaysia Pahang, Malaysia

Prof. Hua Meng, Zhejiang University, China

Dr. Razika Ihaddadene, Med Boudiaf M'Sila University, Algeria

Assoc. Prof. Yasin POLAT,Nevşehir Hacı Bektaş Veli University, Turkey

Assoc. Prof. Pascal Briois, Bourgogne Franche-Comté-UTBM, Montbéliard

Assoc. Prof. D. B. Jani, Government Engineering College, Dahod

Dr. Alfonso Chinnici, University of Adelaide, Australia

Dr. Balaji Trimbak Khogare, Raje Ramrao Mahavidyalaya, India

Dr. Pang - Chieh Sui, Wuhan University of Technology, China

Dr. Nicolae Vaszilcsin,Politehnica University of Timisoara, Romania

Dr. Umair Yaqub Qazi, University of Hafr Al Batin, Saudi Arabia

Dr. Veronica Bermudez, QEERI- Qatar Environment & Energy Research Institute, Qatar

Dr. Gioele Di Marcoberardino, Università degli Studi di Brescia, Italy

Dr. Amandeep Singh Oberoi, Thapar University, India

Dr. Michel Noussan, Politecnico di Torino, Italy

Dr. Francesco Neirotti, Politecnico di Torino, Italy

Dr. Sonil Nanda, University of Saskatchewan, Canada

Dr. Rahat Javaid, Fukushima Renewable Energy Institute - AIST, Japan

Prof. Nejla ALHABIB MAHJOUB SAID, King Khalid University, Kingdom Saudi Arabia

Dr. Miguel Laguna-Bercero, Univeridad de Zaragoza, Spain

Associate Prof. Abdellah BENAMI, Moulay Ismail University of Meknes, Morocco

Prof. Karuppasamy Gurunathan, Alagappa University, India

Prof. Maria Jose Lavorante, Institution of Scientific and Technological Research for Defense (CITEDEF), Argentina

Prof. L.M.Das, Centre For Energy Studies, Indian Institute of Technology Delhi, India

Professor Dr. Ho Soon Min, INTI International University, Malaysia

Assoc. Prof. Yifei Wang, Harbin Institute of Technology, Shenzhen, China

Assoc. Prof.Yi Xue, Xi'an University of Technology, China

Assoc. Prof. T.Maiyalagan, SRM Institute of Science and Technology, India

Assoc. Prof. Mohammad Sadegh Valipour, SemnanUniversity, Iran

Assoc. Prof. Aleksandar Anastasovski, International Balkan University (IBU), Macedonia

Dr. Alban Kuriqi, University of Lisbon, Portugal

Peer Review Statement

All papers published in this volume have been reviewed through pr ocesses administered by the Editors. Reviews were conducted by expert referees to the professional and scientific standards expect ed of a proceedings journal published by IOP Publishing.

Type of peer review: Single anonymous

Conference submission management system: Morressier

Number of submissions received: 31

Number of submissions sent for review: 22

Number of submissions accepted: 13

Acceptance Rate (Submissions Accepted / Submissions Received × 100): 41.9 %

Average number of reviews per paper: 2.60

Total number of reviewers involved: 38

Contact person for queries:

Name: DaisyLee

Affiliation: Organizing Committee of HEET

Email: daisy-lee@heet-18.org

* means value has been edited

Study on the Influence of Hydrogen Blended Transportation on the Safety of Existing X70 Steel in Gas Distribution Pipeline Network

Baoling Guo[a,*] and Yanqi Zhang[a]

[a] Beijing Gas Group Research Institute, Beijing, 100011, China

Corresponding E-mai：guobaoling@bjgas.com

Abstract. Coal to gas is one of the natural gas sources utilized in Beijing, with a hydrogen blend concentration range from 1% to 3% and reaching up to 5% in specific cases. In order to evaluate the safety of transporting hydrogen blended natural gas with X70 pipeline steel in distribution pipelines, this study first conducted the slow strain rate tensile tests under a total pressure of 4 MPa in hydrogen environment. The results show that X70 steel exhibits highly sensitive to a hydrogen blending ratio of 5%. Subsequently, four types of hydrogen compatibility tests were conducted in the above sensitive hydrogen partial pressure include the impact toughness test, fracture toughness test, fatigue crack growth test, and fastening disk pressure test. The findings demonstrated that hydrogen blended natural gas has virtual no effect on the impact and fracture toughness, with only 10% reduction in fatigue crack growth threshold. Finally, based on fatigue life analysis by fracture mechanics evaluation, it can be concluded that the existing natural gas pipe of X70 steel is capable to operate safely over 40 years under a total pressure of 4 MPa with a hydrogen blend concentration of 5%.

1. Introduction

The National Development and Reform Commission and the National Energy Administration issued the Medium-and Long-Term Plan for the Development of Hydrogen Energy Industry (2021-2035) [1] on March 23, 2022, solidifying the position of hydrogen energy as a future pillar of the national energy strategy, and a key carrier for achieving green and low-carbon transformation. The basic rules of hydrogen energy safety are prioritized in the Plan such as hydrogen embrittlement failure, leakage, and diffusion, along with pilot demonstrations of blending hydrogen into the existing natural gas pipelines. Relevant studies and demonstration application have been carried out on hydrogen blended transportation in medium and low-pressure distribution pipelines. Shi found that the Wobbe index and combustion potential of natural gas reduced with hydrogen blend transportation, however the requirement of 12T natural gas for residential users can be achieved under a less than 20% hydrogen blend concentration [2]. Based on the interchangeability calculation, Qiao concluded that the maximum upper limit of hydrogen blend concentration to meet the requirements of 12T natural gas was 23%, while a 10% hydrogen blend concentration would have no significant impact on the safety of residential users [3]. Yang researched the adaptability and safety of using hydrogen blended natural gas in domestic gas appliances based on theoretical analysis and experimental tests, and concluded that the less than 20%

hydrogen blended natural gas could be used in domestic gas appliances without any modification [4]. In the 2004 EU NaturalHy Project, it was found that the fatigue resistance of low-grade steel pipes decreased in case of blending hydrogen, with the fatigue test results indicating that the hydrogen blend concentration should be less than 50% [5]. The above research provided a theoretical basis for the feasibility of hydrogen blended natural gas transportation.

However, the compatibility of materials including hydrogen induced ductility loss, hydrogen blistering and hydrogen induced cracking is the greatest challenges of pure hydrogen and hydrogen blended natural gas transportation.

Hydrogen induced ductility loss is the decrease in material ductility caused by hydrogen dissolving in steel in the atomic or ionic form and accumulating to supersaturation at dislocations and microscopic gaps, which can be minimized by dehydrogenation to prevent pipe embrittlement. The fundamental causes of hydrogen bubbling and hydrogen induced cracks usually considered permanent and irreversible are the gradual buildup of this atomic hydrogen and molecular hydrogen which is compounded by some of this accumulation of atomic hydrogen at the structural defects of the material [6]. Due to the increased likelihood of hydrogen leakage in the pipe [7], the research into pipe safety in case of hydrogen blended transportation has gained international attention.

Shang studied hydrogen induced cracking of low carbon steel that transported hydrogen blended natural gas[8]. The results indicated that the coupling effect of carbon dioxide and hydrogen on materials could significantly hasten the fatigue crack growth rate (FCGR) in hydrogen blended natural gas. In Chaoyang City, Liaoning Province, China, the demonstration project of hydrogen blended natural gas confirmed that 20# steel pipe could meet the safety transmission requirements at of the medium and low pressure [9]. An HSE study and analysis in Britain found that in the pressure range of 0 to 0.2 MPa, a hydrogen blended concentration of less than 20% had no effect on low grade steel pipes in natural gas distribution pipelines [10].

The delivery pressure of natural gas distribution pipelines in Europe and the United States is low, and the pipes are typically made of polyethylene (PE) or low grades steel. China's natural gas distribution pipeline network is among the world's most extensive, with a wide range of working pressures (from 3500 Pa to 6 MPa) and a wide variety of materials (from 20# to Q235B and X70). Hydrogen has been shown to have a greater impact on higher grades pipeline steel, with X80 and X70 being more susceptible to hydrogen induced cracking than X60 steel [11]. Hydrogen blended transportation primarily affects tensile properties and fatigue crack growth performance of pipes. With a design pressure of 4 MPa, X70 steel is primarily used in Beijing's outer ring pipeline network. Hydrogen can cause embrittlement in pipes, which can lead to pipe failure [12-13]. There are currently high-pressure rings, medium-pressure networks, and regional low-pressure connectivity in Beijing's natural gas pipeline network with the complex distribution and near to urban buildings. In case of a gas leak, there is a serious risk to the public safety. It is very important to investigate the sensitive hydrogen partial pressure, the change in mechanical properties and service life of pipes with hydrogen blended transportation.

A sensitive hydrogen partial pressure for X70 steel under a total pressure of 4 MPa was analyzed in this work. Furtherly, four types of mechanical performance tests were conducted based on sensitive hydrogen partial pressure include the impact toughness test, fracture toughness test, fatigue crack growth test, and fastening disk pressure tests. Finally, based on the fracture mechanics method, the critical crack size and fatigue life index of pipe were evaluated, and the service safety of pipes hydrogen blended natural gas transmission was determined.

2. Experimental

2.1. Material and Test Environment

The base material and welds of X70 steel consists of polygonal ferrite (PF), granular bainite (GB) and a small amount of acicular ferrite (AF), as test materials with the microstructure shown in Figure 1. The mass fractions of the main chemical components are C0.09, Si0.35, Mn1.7, P0.015, S0.005, V0.06, Nb0.03, and Ti0.025. The test parameters were selected based on the actual operating conditions of the distribution pipeline network in Beijing Sixth Ring Road, with a maximum operating pressure of 4 MPa and hydrogen blending ratios of 1%, 3%, and 5%. The samples were placed in an autoclave pre-filled with natural gas blended with hydrogen to investigate the impact of different hydrogen blending ratios on the mechanical properties of X70 steel.

Figure 1. Microstructure of X70 Steel.

2.2. Test of Sensitive Hydrogen Partial Pressure

The strength, elongation, reduction of area, and fracture morphology in the samples under room temperature and pressure and different hydrogen partial pressure were compared based on the slow strain rate tensile test method considering the greatest impact on hydrogen embrittlement susceptibility of the material, in order to determine the hydrogen concentration that has the impact on the material properties. The samples were smooth bar-shaped tensile samples according to the ISO 7539-7-2005[14] and ASTM G142-98 (2016) [15], which were sampled along the axial and circumferential direction of the pipeline, with a cross-section diameter of 5 mm. The samples were placed in a normal temperature and pressure environment, and a hydrogen environment that was simulated in an autoclave, with hydrogen blending ratio of 1%, 3%, and 5%. Uniaxial tensile stress was applied at the tensile rate of 1 × 10-6 s-1 until the samples cracked, to obtain the stress-strain curves under different environments.

2.3. Relevant Mechanical Test under Sensitive Hydrogen Partial Pressure

Hydrogen-blended natural gas was prepared based on the selected sensitive hydrogen partial pressure. The components are shown in Table 1. Impact toughness, fracture toughness, fatigue crack growth, and fastening disk pressure tests were conducted to evaluate the impact of hydrogen on the mechanical properties, hydrogen embrittlement susceptibility, and crack growth of materials.

Table 1. Components of hydrogen blended natural gas.

Gas components	**H_2**	**CO_2**	**N_2**	**CH_4**	**C_2H_6**	**Total percentage**
Gas percentage (mol%)	1	0.4067	0.1578	98.4322	0.0033	100
	3	0.1586	0.1601	96.6784	0.0029	100

	5	0.3905	0.1515	94.4547	0.0033	100

2.3.1. Charpy Impact Test. Buried pipelines are subject to impact toughness test at 0 °C according to the ISO 3183-2012[16]. Consequently, the work conducted the Charpy impact test on V-notched samples exposed to 5% hydrogen partial pressure for 30 days at 0 °C, and compared the test results with V-notched samples not exposed to hydrogen at 0 °C to evaluate the impact of hydrogen components on material toughness. The samples had a length of 55 mm, height of 10 mm, and width of 5 mm, with a notch angle of 45° and a notch depth of 2 mm.

2.3.2. Fracture Toughness Test. The crack tip opening displacement (CTOD)test method of materials as specified in ISO 12135-2016[17] was used for the fracture toughness test under normal temperature and pressure, with samples taken along the axial and circumferential direction of the pipeline. For the prefabricated crack process and test method of the samples referring to ISO 7539-9-2017[18]. The samples had a length of 100 mm, width of 20 mm, and thickness of 10 mm, with a notch width of 1.2 mm and an effective crack notch length of 7 mm.

Double cantilever beam (DCB) samples as specified in ISO 7539-6-2011[19] was used for the fracture toughness test in the hydrogen environment. The crack was prefabricated on the samples before the test. The samples had a total width of 120 mm, net width of 110 mm, thickness of 20 mm, with a notch width of 0.7 mm and a notch length of 20 mm. Bolt loading was adopted to enable the crack tip to reach the stress intensity factor, and the sample was then placed in a hydrogen autoclave. The hydrogen embrittlement susceptibility of the material was evaluated based on the crack growth.

2.3.3. Fatigue Crack Growth Test. The fatigue crack growth test was carried out according to ISO 7539-6-2011 and ISO 12108-2012[20]. For the compact tensile C (T) samples that were taken along the axial and circumferential direction of the pipeline. The samples had a width of 70 mm and thickness of 14 mm. The fatigue load that is sine wave loading and frequency 1 Hz was applied to the samples by a testing machine, with maximum load 5.2kN, minimum load 0.52 kN. The tests were conducted at normal temperature and pressure and 5% hydrogen blending ratio.

2.3.4. Fastening Disk Pressure Test. The fastening disk pressure test (disk blasting) was carried out according to the ASTM F1459-06[21]. Different gases were used to pressurize the ambient chamber at a certain pressure rise rate until the sample with a diameter of 58 mm and a thickness of 0.7 mm was blasted after reaching the blasting pressure recorded under different gas environment. Pure hydrogen (99.995%), and 5% hydrogen-blended components were used for the tests.

3. Results & Discussion

3.1. Sensitive Hydrogen Partial Pressure of X70 Steel

To analyze the hydrogen sensitivity of X70 steel base metal and weld joint under a pressure of 4 MPa, a series of slow strain rate tensile tests were conducted by varying the hydrogen partial pressure of 1%, 3%, and 5%. The hydrogen embrittlement susceptibility was evaluated. The results in Figure 2 indicate that different levels of hydrogen partial pressure ratio have virtually no effect on the yield strength and ultimate tensile strength of both the base metal and weld joint. However, there is a slight decrease in the weld elongation of under 5% hydrogen partial pressure.

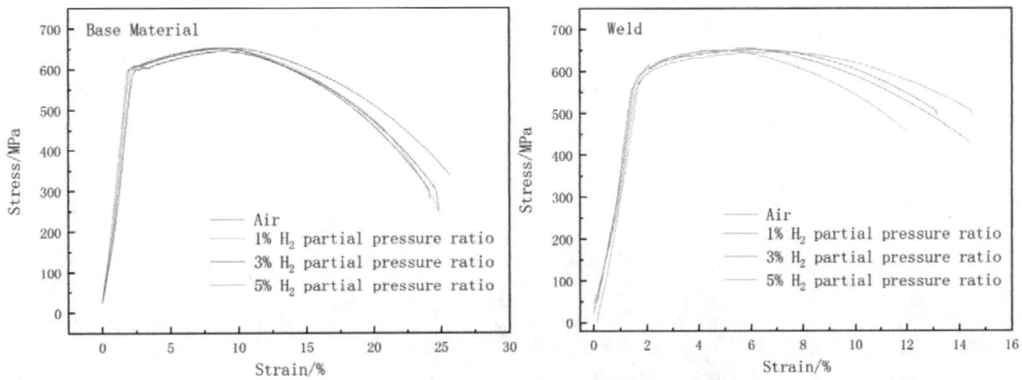

Figure 2. **(a)-(b)** Slow Strain Rate Tensile Curves of X70 Steel Base Material and Welds under Different Hydrogen Partial Pressures.

Hydrogen induced degradation in pipeline steels is typically manifested as a ductility loss of materials. The elongation (δ) and reduction in area (Ψ) of the X70 steel were calculated based on cross-sectional area πd^2 and gauge section length L at initial and final fracture. The calculation formulas for elongation (δ) and reduction of area (Ψ) are shown in Formula (1) and Formula (2), respectively. The embrittlement index (EI) is defined as a measure of hydrogen embrittlement susceptibility, as shown in Formula (3). The higher the reduction of area and elongation, the better the ductility of the material. The fracture morphology in the sample weld is shown in Figure 3.

$$\delta = \frac{L_1 - L_0}{L_0} \times 100\% \quad (1)$$

$$\psi = \frac{\pi d_0^2 - \pi d_1^2}{\pi d_0^2} \times 100\% \quad (2)$$

$$EI = \frac{\psi_{air} - \psi_H}{\psi_{air}} \times 100\% \quad (3)$$

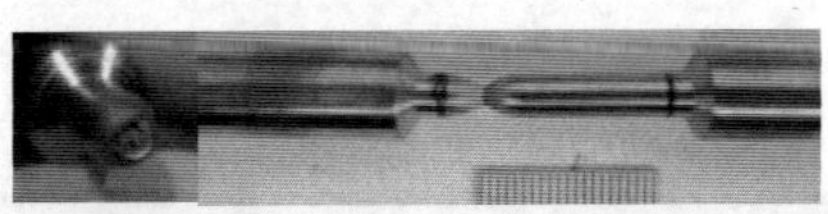

Sample at normal temperature and pressure

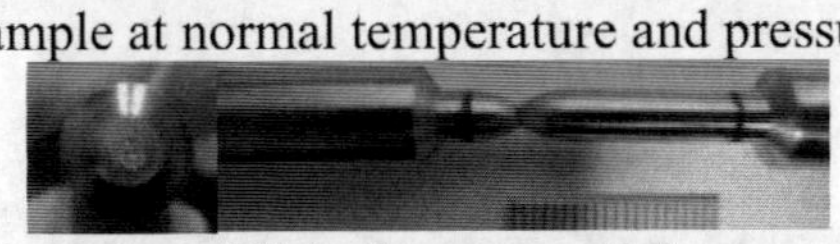

Sample at 1% hydrogen partial pressure

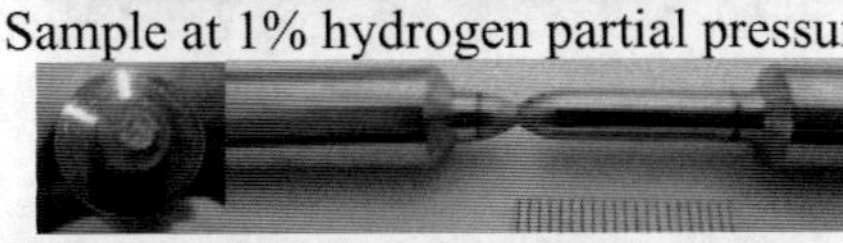

Sample at 3% hydrogen partial pressure

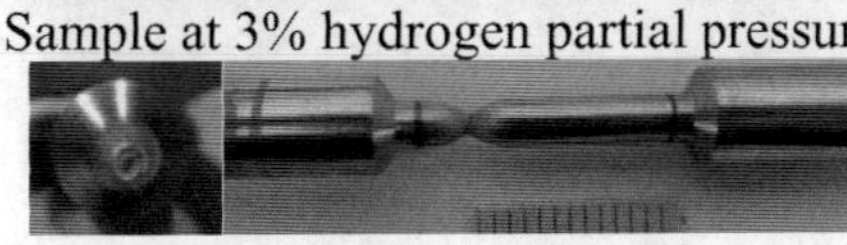

Sample at 5% hydrogen partial pressure

Figure 3. Slow Tensile Fracture Morphology of X70 Steel under Different Hydrogen Partial Pressures.

The same samples tested in the hydrogen environment were compared with those in the non-hydrogen environment to evaluate hydrogen embrittlement susceptibility. The greater the ratio deviation, the higher the cracking sensitivity. It is shown that there is a decrease in the reduction of area and elongation of weld material under 5% hydrogen partial pressure in Figure 4, which indicates that 5% hydrogen partial pressure has an impact on the mechanical properties of the material and increases brittleness of the material. As a result, 5% hydrogen partial pressure is determined as the sensitive hydrogen partial pressure for X70 steel.

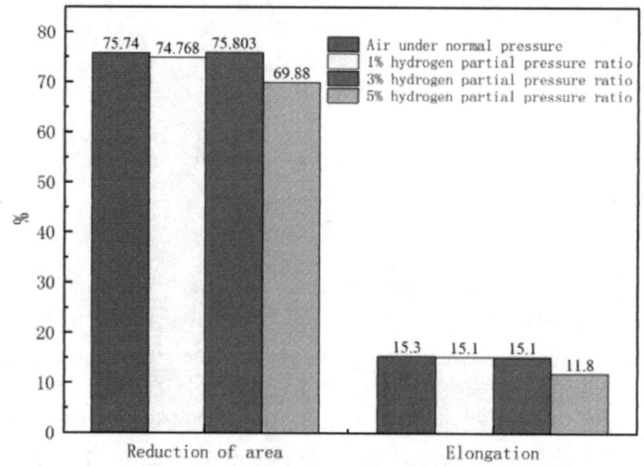

Figure 4. Plasticity Indexes of X70 Steel Welds under Different Hydrogen Partial Pressures.

3.2. Impact of Hydrogen on Relevant Mechanical Properties of X70 Steel under Sensitive Partial Pressure

3.2.1. Impact Toughness Test. The impact toughness test results of X70 steel base material and welds under normal temperature and pressure and 5% hydrogen partial pressure are shown in Table 2.

Table 2. Impact Toughness Test Results of X70 Steel under Normal Temperature and Pressure and 5% Hydrogen Partial Pressure.

Test conditions	**Sample**	**Impact toughness/J**					
		1#	**2#**	**3#**	**4#**	**5#**	**Average**
Normal temperature and pressure	Base material	303	302	310	318	318	310
	Weld	164	174	172	144	165	164
Normal pressure at 0 °C	Base material	282	218	310	276	312	280
	Weld	164	182	170	174	156	169
5% hydrogen partial pressure 30 days, 0 °C	Base material	295	339	335	350	350	334
	Weld	161	181	165	192	159	172

As shown in Figure 5, the impact toughness of samples in a hydrogen environment is not significantly different from that in a non-hydrogen environment, indicating that 5% hydrogen partial pressure has little impact on the impact toughness of materials.

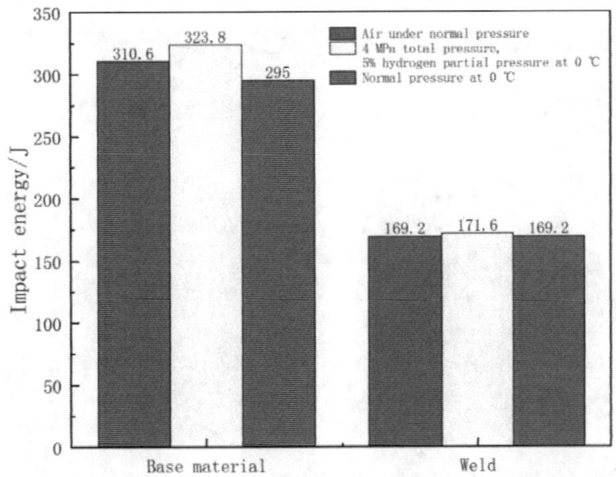

Figure 5. Comparison of Impact Energy Test Results of X70 Steel under Normal Temperature and Pressure and 5% Hydrogen Environment.

3.2.2. Fracture Toughness Test .The tests were conducted on 4 groups of samples, carried out from the transverse and circumferential direction of the pipeline, using DCB samples. The DCB samples were loaded up to the yield strength, with the stress intensity factor at crack tip reaching 25% of the K_{IC} （fracture toughness）. The results, shown in Figure 6 and Figure 7, indicated that the crack tip of the weld and base material samples did not grow forward under 5% hydrogen partial pressure within 30 days and there were no secondary cracks. Consequently, when service stress below the yield strength (e.g., Class 1 natural gas pipelines with a design factor of 0.72), it appears that the blending natural gas of 5% hydrogen has no detrimental effect on the damage tolerance of both X70 parent material and weld joint. This finding holds significant importance in terms of practical applications.

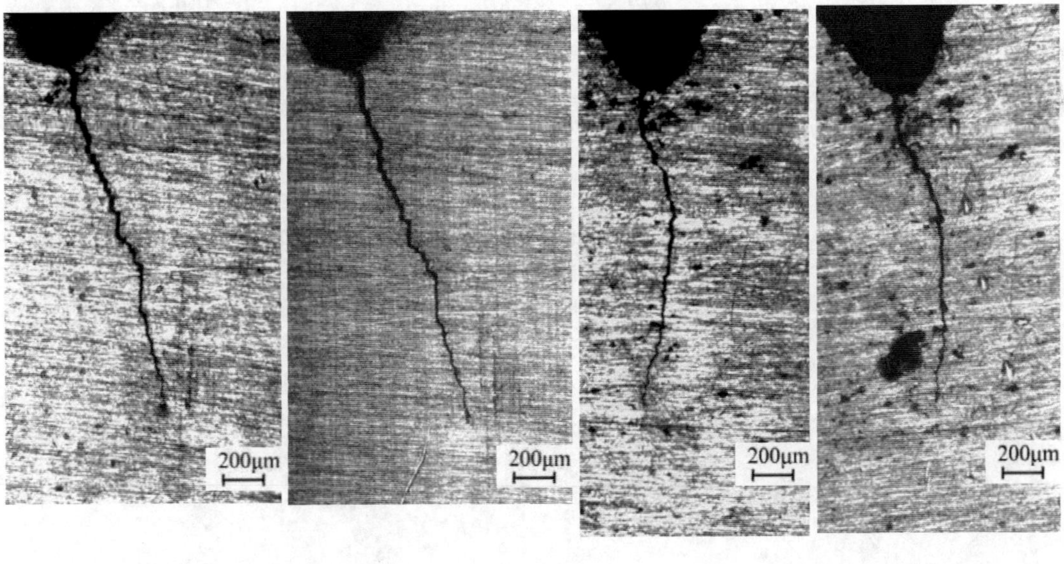

Base metal #1 Base metal # 2

Figure 6. Crack of DCB Samples of X70 Steel Base Material before and after Soaking under 5% Hydrogen Partial Pressure for 30 Days.

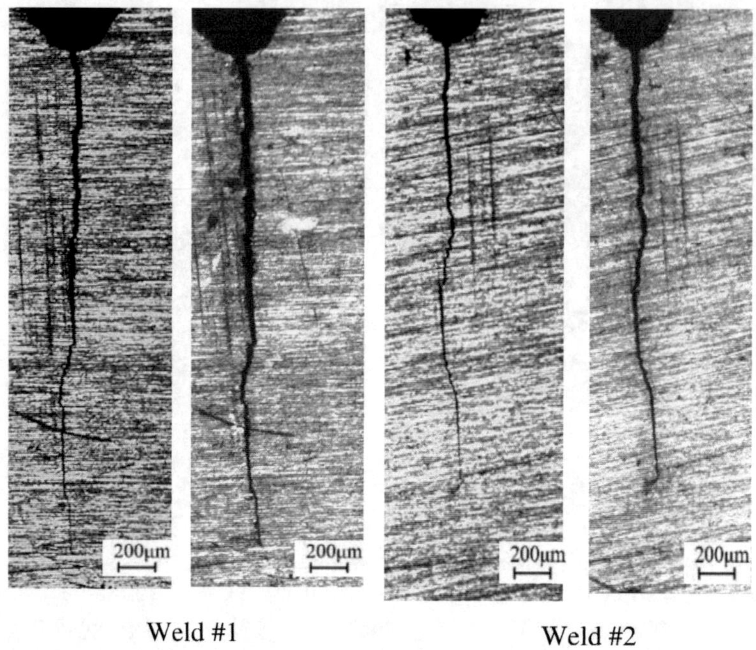

Figure 7. Crack of DCB Samples of X70 Steel Weld before and after Soaking under 5% Hydrogen Partial Pressure Ratio for 30 Days.

3.2.3. Fatigue Crack Growth Test. Hydrogen has a significant impact on the fatigue performance of material, leading to an increase in the fatigue crack growth rate, and a lowering of the threshold of fatigue crack initiation. There effects can occur in combination with each other [22]. If the fatigue crack growth threshold or fatigue crack growth rate of all parallel samples under hydrogen condition is degrade than that under original conditions, then it can be concluded that hydrogen has an effect on X70 pipe steel.The fatigue crack growth rate curve (da/dN v.s. △K) of X70 steel under room RT and 5% hydrogen partial pressure are shown in Figure 8. The fatigue crack growth threshold of the X70 steel base material and the weld sample is lower in the hydrogen environment than that at RT. Table 3 shows the results of determining the fatigue crack growth of X70 steel through linear regression under different environments. The coefficients n and m can be obtained by fitting the Paris' formula,expressed as $\frac{da}{dN} = 10^{-11.23} \Delta K^{5.21}$.

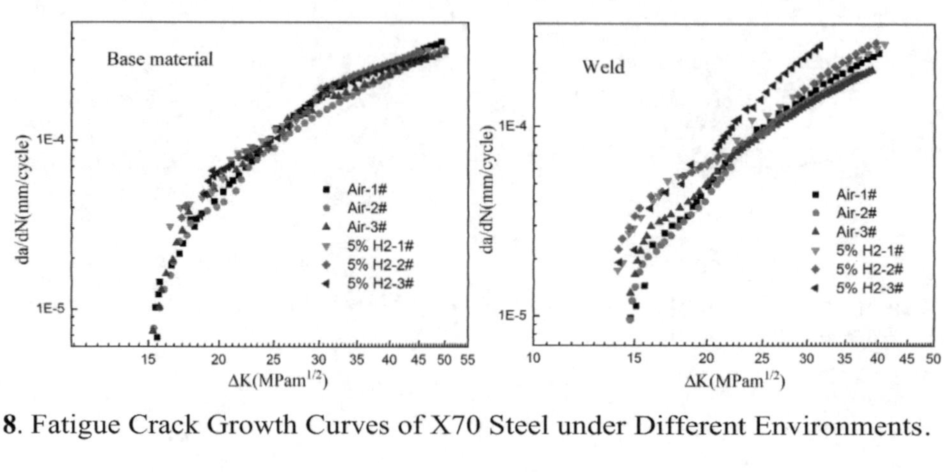

Figure 8. Fatigue Crack Growth Curves of X70 Steel under Different Environments.

Table 3. Amplitude of Fatigue Threshold Stress Intensity Factor of X70 Steel under Normal Temperature and Pressure and 5% Hydrogen Partial Pressure.

Test environment	Sampling type	Fatigue threshold ΔK_{Ith}/MPam$^{1/2}$			
		1#	2#	3#	Average
Air under normal pressure	Base material	15.51	14.92	15.07	15.17
	Weld	14.67	14.80	14.34	14.60
5% hydrogen partial pressure	Base material	13.14	13.67	14.01	13.60
	Weld	13.76	13.54	12.03	13.11

The threshold for fatigue crack growth of the two samples is slightly less than 15 MPam1/2 as shown in table 3, with a reduction of 10% when blending 5% hydrogen. It is evident that the △Kth of all parallel samples in the hydrogen environment are lower than those at RT. Therefore, there is a decrease in the fatigue properties of X70 steel under 5% hydrogen partial pressure, especially in the weld joint. Even a small amount of hydrogen can adversely affect the fatigue performance of pipeline steel, which is consistent with the findings of Meng's research. (Hydrogen effects on X80 pipeline steel in high-pressure natural gas/hydrogen mixtures)

3.2.4. Fastening Disk Pressure Tests (Disk Blasting). A total of 15 groups of fastening disk pressure tests were carried out on X70 steel base material including 6 groups in a helium environment and 9 groups in a hydrogen environment. The test results are shown in Table 4. The blasting pressure of X70 steel in the hydrogen environment is slightly lower than that in the helium environment, indicating an increase in material brittleness. The blasting pressure ratio PHe/PH2 in the helium and hydrogen environment is shown in Table 5. long-term exposure to hydrogen may result in hydrogen embrittlement at when PHe/PH2 is greater than 1 but less than 2. Consequently, based on the analysis of the test results, the material has certain hydrogen embrittlement susceptibility long-term exposure to hydrogen environment.

Table 4. Disk Blasting Test Results of X70 Steel under Different Conditions.

Experiment conditions	Pressurization rate bar/min					
	1	5	10	50	100	500
Helium/MPa	30.91	33.31	33.58	33.15	33.14	32.14
Hydrogen/MPa	31.40	33.01	32.48	32.12	33.05	29.53

Table 5. Evaluation of Experimental Indexes in the Fastening Disk (Disk Blasting) Pressure Test.

Evaluation	P_{He}/P_{H2} test results					
	1	5	10	50	100	500
P_{He}/P_{H2}=1, the material is insensitive to hydrogen; $P_{He}/P_{H2} > 2$, the material is sensitive to hydrogen 2> $P_{He}/P_{H2} < 1$, long-term exposure to hydrogen environment may lead to hydrogen embrittlement	0.98	1.01	1.03	1.03	1.00	1.09

3.3. Pipe Fitness for Service Assessment

Given the potentially accelerated fatigue failure of natural gas pipes during hydrogen blended transportation, this work employed the Paris' formula to calculate the fatigue crack growth rate, expressed as $\frac{da}{dN} = C\Delta K^{n}$. The stress intensity factor range (ΔK) was determined based on the test data under the operation conditions of Beijing's urban gas pipeline network. The formula for calculating ΔK is $\Delta K = \frac{\Delta P}{B\sqrt{W}} \times \frac{(2+\alpha)}{(1-\alpha)^{3/2}} \times \left(0.886 + 4.64\alpha - 13.32\alpha^2 + 14.72\alpha^3 - 5.6\alpha^4\right)$.The formula $K_{\mathrm{mat}} = min\left[8.47(CVN)^{0.63}, 14.6(CVN)^{0.5}\right]$ was used to calculate the initial crack depth a_0 and fracture toughness (K_{mat}) of the material under the hydrogen blended environment according to the KHKS 0220[23] and ISO 12135-2016. Finally, iterative calculation was conducted until the maximum stress intensity factor K_{max} was exceeded than or equal to the fracture toughness K_{mat}, in order to obtain the critical crack size (a_1) and fatigue life (N_{P}) of the hydrogen blended natural gas pipeline to judge the fatigue failure of the material. The detailed process is illustrated in Figure 9.

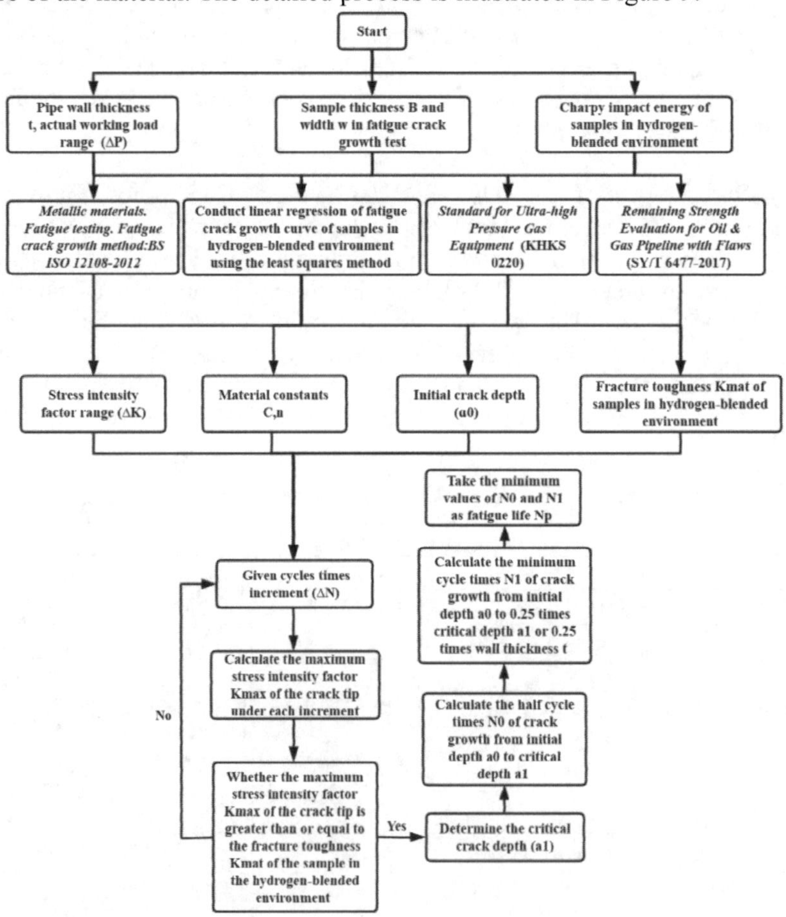

Figure 9. Flowchart of Material Failure Assessment.

Currently, in Beijing's urban gas pipeline network, X70 steel is used with an outer diameter of 1m, with wall thickness of not less than 8.7 mm and a design pressure of 4 MPa. Through iterative calculation, the fatigue cycle times of in the hydrogen blended environment exceeds 5,000, with a critical crack depth of 6.1 mm. These results indicate the X70 steel can safely operate for over 40 years to transport hydrogen blended natural gas at a total pressure of 4 MPa with a hydrogen blend concentration of 5%.

4. Conclusion

a. The results based on slow strain tensile test, it was found that the sensitive hydrogen partial pressure ration for X70 steel pipe is 5%.

b. The results from Charpy impact and DCB (Double Cantilever Beam) tests show that the presence of 5% hydrogen in natural gas does not affect the toughness of X70 base material and welds. Under the condition of applying a 25% of K_{IC} stress intensity factor, the X70 steel demonstrates the ability to resist crack propagation in hydrogen environment.

c. The results of fatigue crack growth rate indicate that the addition of 5% hydrogen reduces the fatigue crack growth threshold of X70 base material and welds by approximately 10%. Additionally, the fatigue crack growth rate is significantly increased. Fatigue life evaluation was conducted using fracture mechanics assessment methods, and the results demonstrate that X70 steel can safely operate for over 40 years under a total pressure of 4 MPa and a hydrogen blend concentration of 5%.

References

[1] National Development and Reform Commission and National Energy Administration 2022 Medium-and long-term plan for the development of hydrogen energy industry (2021-2035) issued *North China Power* **14**(03), 4

[2] Shi H, Lv Y and Tan G B 2022 Feasibility study on gas pipeline hydrogen-mixing transportation *Natural Gas and Oil* **40**(04), 23-31

[3] Qiao J, Yan S, Guo B L, Ma X Q and Cheng W H 2022 Influence of hcng on city gas users *Gas & Heat* **42**(08), 36-39

[4] Yang M C, et al. 2022 Evaluation on applicability of household gas appliances to hcng and their demonstrative application *Mechanics in Engineering* **44**(04), 794-800

[5] Tiekstra G 2008 The naturalhy project: first step in assessing the potential of the existing natural gas network for hydrogen delivery *Lunión Médicale Du Canada* **114**(03), 213-219

[6] Feng H, Chi Q, Ji L K, Li H and Yang K 2017 Research and development of hydrogen embrittlement of pipeline steel *Corros. Sci. Prot. Technol* **29** (03), 318-322

[7] Yu Z L, Zhang L Y, Ning C, Sun G, Liu H F, An Z H, Wang J H, Huang Z H, Lu Y H and Yang M C 2022 Hcng pipeline transportation and terminal application *Mechanics in Engineering* **44**(03), 491-502

[8] Shang J, Chen W F, Zheng J Y, Hua Z L, Zhang, L, Zhou C S and Gu C H 2020 Enhanced hydrogen embrittlement of low-carbon steel to natural gas/hydrogen mixtures *Scr. Mater.* **189**(01), 67-71

[9] Zhong B, Zhang X X, Zhang B and Peng S P 2022 Research on development of hcng industry in china *Strategic Study of CAE* **24**(03), 100-107

[10] Birkitt K, Loo-Morrey M, Sanchez C and O'Sullivan L 2020 Materials aspects associated with the addition of up to 20 mol% hydrogen into an existing natural gas distribution network *Int. J. Hydrog. Energy* **46**(23), 12290-12299

[11] Ren R X, You S J, Zhu X Y, Yue X W and Jiang Z C 2021 Development Status and Prospects of HCNG Transportation Technology *Petroleum and New Energy* **33**(04), 26-32

[12] Rajabipour A and Melchers R E 2018 Service life of corrosion pitted pipes subject to fatigue loading and hydrogen embrittlement *Int. J. Hydrog. Energy* **43**(17), 8440-8450

[13] Stenerud G, Wenner S, Olsen J S and Johnsen R 2018 Effect of different microstructural features on the hydrogen embrittlement susceptibility of alloy 718 *Int. J. Hydrog. Energy* **43**(13), 6765-6776

[14] Corrosion of metals and alloys. Stress corrosion testing. Part 7: Method for slow strain rate testing ISO 7539-7: 2005: EN ISO 7539-7-2005.

[15] Standard Test Method for Determination of Susceptibility of Metals to Embrittlement in Hydrogen Containing Environments at High Pressure, High Temperature, or Both: ASTM

G142-98(2016). 2016.
[16] Draft Document - Petroleum and natural gas industries - Steel pipe for pipeline transportation systems (ISO 3183:2012/DAM 1:2016); German and English version EN ISO 3183:2012/prA1:2016: DIN EN ISO 3183/A1 - DRAFT. 2016.
[17] Metallic materials -- Unified method of test for the determination of quasistatic fracture toughness:ISO 12135-2016. 2016.
[18] Corrosion of metals and alloys - Stress corrosion testing - Part 9: Preparation and use pre-cracked specimens for tests under rising load or rising displacement:KS D ISO 7539-9-2017. 2017.
[19] Corrosion of metals and alloys. Stress corrosion testing. Part 6: Preparation and use of precracked specimens for tests under constant load or constant displacement:BS EN ISO 7539-6-2011.
[20] Metallic materials. Fatigue testing. Fatigue crack growth method:BS ISO 12108-2012. 2012.
[21] Standard Test Method for Determination of the Susceptibility of Metallic Materials to Hydrogen Gas Embrittlement (HGE): ASTM F1459-06(2017). 2017.
[22] Lam P S, Sindelar R L, Duncan A J, Adams T M 2009 Literature Survey of Gaseous Hydrogen Effects on the Mechanical Properties of Carbon and Low Alloy Steels *J Press Vessel Technol* **131**(04), 041408-1-041408
[23] Japan High Pressure Technology Association. Standard for ultra-high pressure gas equipment. KHKS 0220, 2020:38.

Techno-economic analysis of hydrogen production systems based on solar/wind hybrid renewable energy

Chunyan Song[1], Bo Zhang[1], Wenchao Cai[2], Dounan Zhang[3,*], Xianglin Yan[4], Tianqi Yang[4] and Jinsheng Xiao[4]

[1] Shandong Electric Power Engineering Consulting Institute Corp. Ltd., Jinan 250061, China

[2] VOYAH Automobile Technology Co., Ltd, Wuhan, Hubei 430050, China

[3] Strategic Account Department, Dongfeng Usharing Technology Co., Ltd., Wuhan 430010, China

[4] School of Automotive Engineering, Wuhan University of Technology, Wuhan 430070, China

2425r26hd@163.com

Abstract. With the popularity of renewable energy, more and more countries and regions are utilizing renewable energy to produce hydrogen. However, renewable energy hydrogen production systems are often large in capacity and equipment, and require rational design as well as performance analysis of the system to satisfy the technicalities while ensuring the economics. In this paper, 12 different renewable energy systems for hydrogen production, including 6 grid-connected systems and 6 off-grid systems, are designed considering solar and wind energy as the main renewable energy sources in a place in China. Subsequently, simulations were carried out using HOMER software to compare and analyze the economic and technical performance of different schemes of renewable energy hydrogen production systems, and the best grid-connected and off-grid renewable energy hydrogen production system schemes were selected based on the comparison results.

1. Introduction

Renewable energy sources such as solar and wind are widely used internationally, but their fluctuating and intermittent nature leads to the need for the use of energy storage technologies to store the energy generated by the system in a timely manner [1]. Currently, the energy storage technology that is in line with the concept of green development is hydrogen energy storage technology. Renewable energy hydrogen production technology utilizes electricity generated from renewable energy power generators for electrolysis of water to produce hydrogen, realizing the conversion from electricity to hydrogen. Since renewable energy hydrogen production systems usually have large component capacities and high construction costs, it is necessary to rationally design the system components and system capacity so that the system can meet the technical performance while ensuring the economy.

In order to rationally design the renewable energy hydrogen production system (HPS), many scholars have carried out simulation studies of the system. By modeling different types of renewable energy HPSs, scholars study the relationship between different input parameters (light intensity, wind speed, etc.) and key output parameters such as power generation and hydrogen production of the system, and analyze the net present value (NPV) of the system, energy cost and other key indicators. Di Micco et al. [2] conducted a techno-economic analysis of two Renewable Multi-Energy Systems and showed that the multi-energy cogeneration system can effectively reduce the averaging cost, which is 0.048-0.054 $/kWh for electricity and 3.14-3.49 $/kg for hydrogen. Based on the weather data in Northwest Germany, Schnuelle et al. [3] built an off-grid photovoltaic (PV)/wind turbine (WT)/fuel cell (FC) hydrogen production system in order to analyze the output performance of the system. The results show that the system's hydrogen production efficiency, renewable energy utilization, hydrogen yield, and energy cost are highly dependent on the system's output power. Based on weather data in the Mangir region of Iran, an off-grid PV/WT hydrogen production system was established by Rezaei et al [4]. It was found that the system produces about 91 kg of hydrogen per day in the Manjir region and saves about 1,347 liters of gasoline in a week. Marocco et al. [5] used weather data from Italy, Greece and Norway to set up a renewable energy HPS with different scenarios to analyze the system techno-economically. According to the results, the HPS with an energy storage device is less dependent on the external supply of electricity. He et al. [6] analyzed the feasibility of establishing renewable energy systems on remote islands in Japan. The results show that a system including an electrolyzer can effectively reduce the system cost by 16.6%. Based on the weather data of Izmir, Istanbul and Ankara regions of Turkey, Karacavus et al. [7] analyzed the power generation and hydrogen production of the PV hydrogen production systems in the three regions using TRNSYS and GenOpt software. The results showed that the total power generation in the three regions was 42803.66kWh, 42573.74kWh, and 44613.95kWh in one year's time; and the hydrogen production was 10488.39m3, 9824.70m3, and 10368.65m3, respectively. Based on the weather data of Hami region in Xinjiang, Fan et al. [8] established an off-grid PV/WT/FC hydrogen production system and performed a techno-economic analysis of the system. The results showed that the system can fully utilize local renewable energy sources such as wind and solar energy, and reduce the environmental pollution caused by traditional power generation methods.

In summary, the simulation of renewable energy HPSs has been studied by many scholars, but most of the research cases are only for a single type of renewable energy HPS, and there is no comprehensive consideration of different types of systems for different scenarios. In addition, many studies have focused on optimizing the performance of the system as well as the capacity configuration, while neglecting the applicability of the system itself, making it difficult to come up with a renewable hydrogen system that is most suitable for local development. In this paper, different schemes of grid-connected and off-grid renewable energy HPSs are designed by combining solar and wind energy with a certain place in China as the research site, and the HPSs of different schemes are comprehensively compared and analyzed from the economic and technical perspectives in order to select the optimal grid-connected and off-grid HPSs.

2. System component model

Renewable hydrogen production systems are feature-rich, providing both electricity and hydrogen, and consist mainly of power generation components (PV panels, WT, etc.), electrolyzers, hydrogen storage tanks, FC, and the user end (electricity users, hydrogen users, etc.).

In this paper, monocrystalline silicon PV cells are used. In practical engineering applications, the PV cell output current I_{pv} and voltage U_{pv} are usually calculated by the empirical equations shown below [9]:

$$\begin{cases} I_{pv} = I_{sc} - I_{sc}C_1\left(e^{\frac{U_{pv}}{C_2 U_{oc}}} - 1\right) \\ C_1 = \left(1 - \frac{I_m}{I_{sc}}\right)e^{\left(-\frac{U_m}{C_2 U_{oc}}\right)} \\ C_2 = \left(\frac{U_m}{U_{oc}} - 1\right)/\ln\left(1 - \frac{I_m}{I_{sc}}\right) \end{cases} \tag{1}$$

where, I_{sc} is the short-circuit current, A; U_{oc} is the open-circuit voltage, V; I_m is the maximum power point current, A; U_m is the maximum power point voltage, V.

The relationship between the output power of the WT and the wind speed can be expressed using a segmented function with the expression:

$$P_W = \begin{cases} 0, & 0 \le v < v_1; v > v_2 \\ \frac{1}{2}\rho\pi R^2 C_p v^3, & v_1 \le v < v_r \\ P_r, & v_r \le v \le v_2 \end{cases} \tag{2}$$

where v_1, v_r and v_2 are the cut-in wind speed, rated wind speed and cut-out wind speed of the WT, m/s, respectively. When the wind speed is less than the cut-in wind speed or more than the cut-out wind speed, the WT will not work. P_r is the rated power of the WT, W.

The hydrogen production rate of an alkaline electrolyzer can be calculated by obtaining the U-I characteristic curve of the alkaline electrolyzer under normal operating conditions. The expression for the rate of hydrogen production is:

$$q_{H_2} = n\eta_F \frac{I_e}{2F} \tag{3}$$

where q_{H_2} is the rate of hydrogen production, mol/s; F is the Faraday constant with a value of 96485C/mol; I_e is the electrolyzer operating current, A; η_F is the electrolyzer current efficiency. Based on practical engineering experience, η_F is expressed as:

$$\eta_F = \frac{(I_e/A_e)^2}{k_{f1}+(I_e/A_e)^2}k_{f2} \tag{4}$$

where A_e is the surface area of the electrode plate of the electrolyzer, m^2; k_{f1} and k_{f2} are both current efficiency parameters of the electrolyzer.

A proton exchange membrane fuel cell is composed of n_{fc} individual cells connected in series and its output voltage U_{fc} can be expressed as:

$$U_{fc} = n_{fc}(E - U_{act} - U_{ohm} - U_{conc}) \tag{5}$$

where E, U_{act}, U_{ohm}, and U_{conc} are the thermodynamic potential, activation overvoltage, ohmic overvoltage, and concentration overvoltage, V, respectively.

3. Renewable Energy Hydrogen Production System Program Design and Parameterization

In this section, a place in China is taken as the study site, and based on the distribution characteristics of renewable energy in this place, 12 different scenarios of renewable energy HPSs are established, including six grid-connected systems and six off-grid systems.

The basic structure of different schemes of renewable energy HPS is shown in figure 1. Renewable energy HPSs are categorized into grid-connected and off-grid systems based on whether they are connected to the grid or not, including: grid-connected/off-grid PV hydrogen production systems (G1/S1),grid-connected/off-grid PV/FC hydrogen production system (G2/S2), grid-connected/off-grid WT hydrogen production system (G3/S3), grid-connected/off-grid WT/FC hydrogen production system (G4/S4), grid-connected/off-grid PV/WT hydrogen production system (G5/S5), and grid-connected/off-grid PV/WT/FC hydrogen production system (G6/S6).

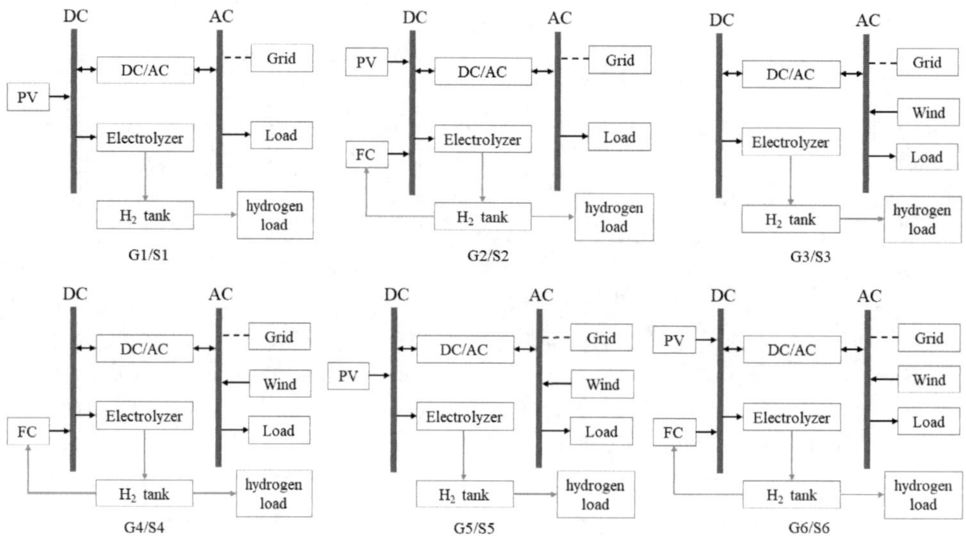

Figure 1. Schematic illustration of the basic structure of renewable energy hydrogen production systems for different scenarios.

For a grid-connected system, when the power generation of the system is less than the electrical load, the system takes power from the grid to meet the electrical load and hydrogen load, and when the power generation of the system meets the electrical load and hydrogen load, the remaining power is delivered to the grid. For off-grid renewable energy HPS, when the power generation of the system is greater than the electric load, the excess electricity will be used by the electrolyzer to produce hydrogen. When the power generation of the system is less than the electrical load, the hydrogen in the hydrogen storage tank can be used to generate electricity for the electrical load for the system configured with FCs components.

The input parameters for the simulation need to take into account weather data, load (electrical and hydrogen load) data, and system component parameters. Weather data are based on data compiled by the National Aeronautics and Space Administration. In this paper, the average monthly solar radiation intensity and clearness indices are used as the main basis for the evaluation of solar energy potential, and the average monthly wind speed is the main basis for the evaluation of wind energy potential. The average solar radiation intensity and clearness indices for different months in a year are shown in figure 2 (a). The annual average clear sky index is 0.572, and the largest and smallest monthly average clear sky index are January and

June, with values of 0.654 and 0.535, respectively, indicating that the proportion of sunny days throughout the year is high and the region is sunny. In addition, the average annual solar radiation intensity in the region is 4.311kWh/m^2/month, and the maximum and minimum monthly average radiation intensity are 6.205kWh/m^2/month and 2.028kWh/m^2/month in June and December, respectively. The monthly average data of wind speed and temperature are shown in figure 2(b). The annual average wind speed is 3.758m/s, the maximum and minimum monthly average wind speed are 5.020m/s and 2.110m/s in May and December, respectively, and the wind speed fluctuates significantly throughout the year. The annual average temperature is 10.19 °C, and the maximum and minimum monthly average temperatures are 25.19 °C and -7.55 °C in July and January, respectively.

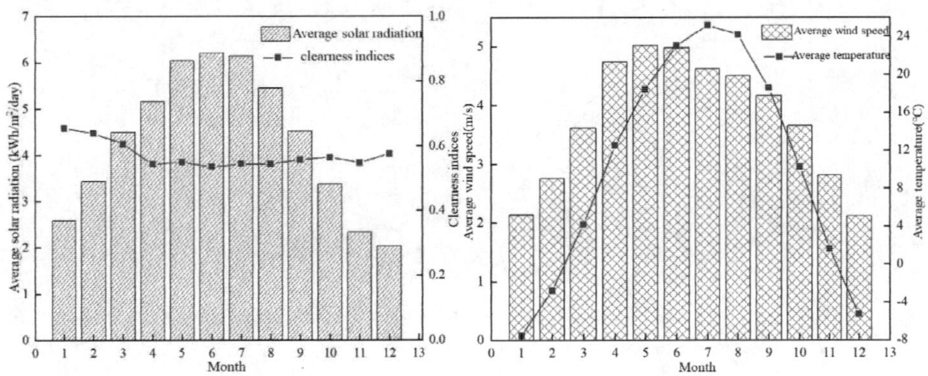

Figure 2. Weather data for the region: (a) monthly averages of solar radiation intensity and clearness indices; (b) monthly averages of wind speed and temperature.

The electrical load data comes from the average monthly electricity demand of a small office building from 0-23 hours provided by HOMER software. As shown in figure 3, electrical load demand is highest in the month of August and the peak time period for electrical load demand occurs from 8-16 pm each day. The average daily electrical load is 172kWh and the average hourly electrical load is 7.2kWh. The hydrogen loading data was obtained from the literature [10], as shown in figure 4, the highest hydrogen loading of 4 kg per hour was observed at 8am -9pm. The daily hydrogen loading was 75 kg, with an average hydrogen loading of 3.13 kg per hour.

The economic parameters of each component of the system are shown in Table 1. The PV panels were selected as general-purpose flat panels, setting the upper power limit to 1200kW and the lower limit to 0. Considering that the minimum monthly average wind speed in the area is about 2m/s, the model selected for the WT is Enercon, with five power ratings including 500kW, 660kW, 800kW, 1000kW, and 2000kW. The electrolyzer was selected as a general-purpose electrolyzer, and the upper power limit was set to 1000 kW, the lower limit to 0, and the working efficiency was set to 85%. The capacity of the hydrogen storage tank is set to an upper limit of 1000 kg and a lower limit of 0. The initial hydrogen mass in the hydrogen storage tank is set to 10% of the capacity of the tank. The fuel cell is set to have an upper power limit of 100kW and a lower limit of 0. The converter is set to have an upper power limit of 1,000kW and a lower limit of 0. Both the inverter and the rectifier operate at 95% efficiency. In addition, for grid-connected renewable HPSs, the cost of the grid is not considered, and the actual role of the grid is equivalent to an infinitely large energy storage device

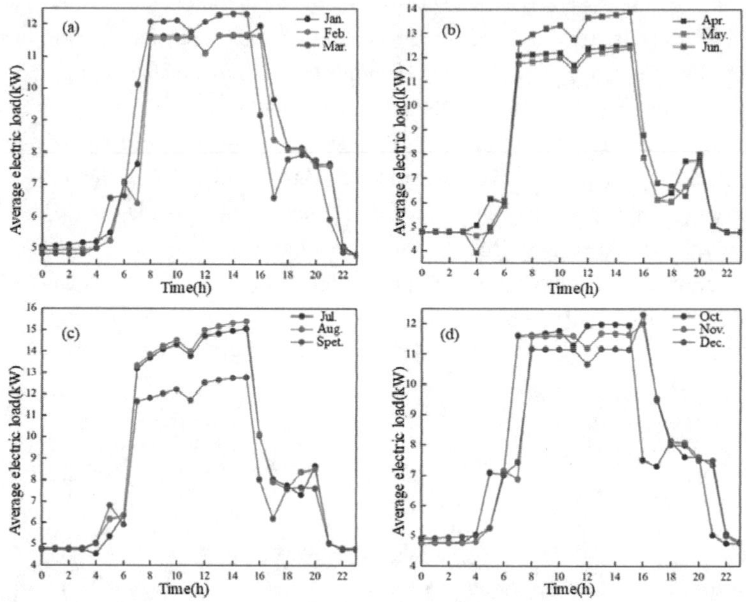

Figure 3. Average electrical load data for a small office building from 0-23 hours per month for one year.

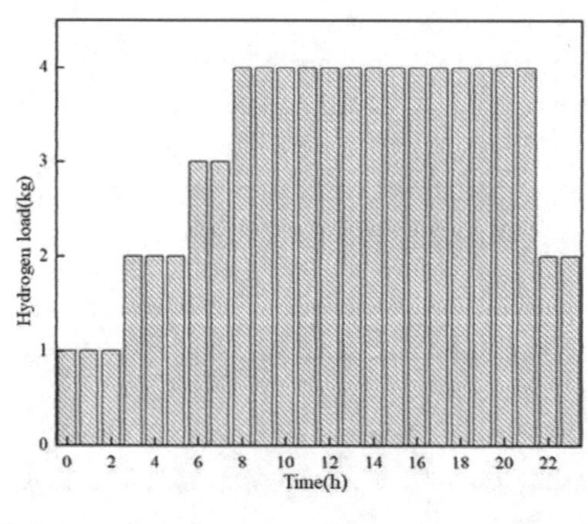

Figure 4. Daily hydrogen loading data from 0-23 hours.

Table 1. System component parameters.

system component	Initial investment cost	Replacement cost	Operation and maintenance costs	life cycle
PV panel	2000$/kW	2000$/kW	10$/kW	30 years
wind turbine	1000$/kW	1000$/kW	12$/kW	30 years
electrolyzer	2000$/kW	2000$/kW	2$/kW	15 years
H_2 storage tank	600$/kg	400$/kg	3$/kg	25 years
Fuel cell	1200$/kW	1200$/kW	0.02$/h	40000 hours

Convertor	700$/kW	630$/kW	10$/kW	15 years

4. Renewable Energy Hydrogen Production System Performance Analysis and Program Selection

4.1 Economic analysis

In this paper, HOMER software is used to simulate and analyze renewable energy HPSs for 12 off-grid and grid-connected scenarios. Firstly, with the objective of minimizing the NPV of the system, the capacity configurations of different scenarios to satisfy the electrical and hydrogen loads of the system are calculated and the results are shown in Table 2. Among them, scenarios S1, S3, and S5 are infeasible, and even an infinite increase in the rated power of the system components will not satisfy the load demand of the system. The main reason for this is that the fluctuating and intermittent nature of solar and wind energy results in systems that do not generate electricity at all times, leading to excessive capacity shortages when the system is not generating enough electricity.

Table 2. Capacity configuration of renewable energy hydrogen production systems for different scenarios at minimum NPV.

Scenario	PV panel (kW)	Wind turbine (kW)	Electrolyzer (kW)	H_2 storage tank (kg)	Fuel cell (kW)	Convertor (kW)
G1	900	/	500	300	/	10
G2	900	/	600	250	10	10
G3	/	4×1000	350	400	/	400
G4	/	4×1000	400	500	5	400
G5	800	800	500	300	/	100
G6	800	800	600	300	5	100
S2	900	/	600	250	10	20
S4	/	4×1000	400	500	20	400
S6	800	800	600	300	10	100

After calculating the appropriate capacity configurations, a comparative analysis of the different options of grid-connected and off-grid renewable energy HPSs will be performed in terms of system economics and technicality, respectively. The economic analysis is to compare and analyze the investment cost, operation cost, NPV, and energy cost of different scenarios respectively. According to the simulation results of HOMER software, the investment cost, operating cost, salvage value, NPV, and energy cost of the grid-connected and off-grid renewable energy hydrogen production system for different scenarios are shown in Table 3. It can be seen that among the grid-connected and off-grid renewable energy systems for hydrogen production, the investment cost, operating cost, NPV and energy cost are smaller for the systems containing PV panels (G1, G2, G5, G6, S2, S6), while they are larger for the systems that do not contain PV panels (G3, G4, S4). The main reason for this is that solar energy is more abundant than wind energy at this location, and when PV panels and WT produce the same amount of electricity, the WT is larger and the system is more costly, while at the same time the difference in actual electricity consumed by the electric load is not significant, resulting in excessive energy costs for the system. In addition, the hydrogen production systems without FCs (G1, G3, G5) have smaller investment costs, operating costs, and NPVs than the hydrogen production systems with FCs (G2, G4, G6), provided that the other components of the grid-connected hydrogen production systems are the same. Grid-connected (G2, G4, G6) have a smaller NPV than off-grid (S2, S4, S6) hydrogen production systems with the same components of the hydrogen production system, mainly due to the fact that the grid-connected hydrogen production system can deliver excess power to the grid to increase the system revenue.

In summary, analyzing from the perspectives of investment cost, operation cost, NPV and energy cost, for the grid-connected renewable energy HPS, the PV hydrogen production system (G1) and the PV/WT hydrogen production system (G5) are economically better; for the off-grid renewable energy hydrogen production system, the PV/FC hydrogen production system (S2) and the PV/WT/FC system (S6) are economically better.

Table 3. Data on economic indicators of renewable energy hydrogen production systems for different scenarios.

Scenario	Investment cost (10^6\$)	Operating cost (10^4\$/year)	Salvage value (10^6\$)	Net present value (10^6\$)	Energy cost (\$/kWh)
G1	3.92	3.12	0.64	3.59	0.162
G2	3.96	3.25	0.71	3.62	0.174
G3	5.26	8.12	1.02	6.06	0.268
G4	5.39	8.24	1.07	6.12	0.273
G5	4.05	4.76	1.29	4.30	0.152
G6	4.26	4.95	1.31	4.38	0.156
S2	4.02	3.29	0.71	3.98	0.192
S4	5.40	8.26	1.07	6.32	0.284
S6	4.27	5.01	1.31	4.42	0.162

4.2 Technical analysis

Technical analysis typically analyzes the output performance of a system over its full life cycle. This section compares and analyzes the generation, hydrogen production, hydrogen use, capacity shortage rate, and excess power rate (off-grid) of HPS for different scenarios. The power generation of the renewable HPS for different scenarios is shown in Table 4. For the grid-connected system, it can be seen that the total power generation of scenarios G3 and G4 is significantly larger than the other scenarios. Meanwhile, it can be found that the hydrogen production system configured with FCs has a larger total power and a smaller grid-supplied power than the hydrogen production system without FCs, when the other components of the system are the same. For the off-grid system, the total generation of Scenarios S4 and S6 is significantly higher than the total generation of Scenario S2. Therefore, from the point of view of system power generation, WT/FC (G4) and PV/WT/FC (G6) hydrogen production systems are technically better for the grid-connected type of hydrogen production system, and WT/FC (S4) and PV/WT/FC (S6) hydrogen production systems are technically better for the off-grid type of hydrogen production system.

Table 4. Electricity generation from renewable hydrogen production systems for different scenarios.

Scenario	PV power generation	WT power generation	FC power generation	Grid generation (kWh/year)	Total power (kWh/year))

	(kWh/year)	(kWh/year)	(kWh/year)		
G1	1,513,124	/	/	30,430	1,543,554
G2	1,513,124	/	38,935	9,517	1,561,576
G3	/	3,124,942	/	14,563	3,139,505
G4	/	3,124,942	40,489	6,267	3,171,698
G5	1,194,546	926,190	/	7,013	2,127,749
G6	1,194,546	926,190	30,135	2,523	2,153,394
S2	1,513,124	/	22,522	/	1,535,646
S4	/	3,124,942	12,288	/	3,137,230
S6	1,194,546	926,190	5,478	/	2,126,214

The amount of hydrogen produced and used by the renewable energy hydrogen production system for different scenarios is shown in Table 5. It can be seen that for the grid-connected hydrogen production system, the hydrogen production system equipped with a FCs (G2, G4 G6) produces more hydrogen and uses less hydrogen at the same time than the hydrogen production system without FCs (G1, G3 G5), provided that the other components of the system are the same. This is mainly due to the fact that the fuel cell utilizes some of the hydrogen to generate electricity, resulting in a reduction in the actual amount of hydrogen used by the user. In the off-grid type hydrogen production system, S2 produces more hydrogen than S4 and S6 hydrogen production systems, but at the same time the fuel cell uses more hydrogen. For Scenario S6, although the amount of hydrogen produced is smaller, the amount of hydrogen used by the fuel cell is also smaller, so the actual amount of hydrogen used by the user is larger. Therefore, from the perspective of system hydrogen production and consumption, for the grid-connected hydrogen production system, the G3 and G5 hydrogen production systems are technically better in terms of larger actual hydrogen consumption by users. For the off-grid hydrogen production system, the S2 and S6 hydrogen production systems are technically better.

Table 5. Hydrogen production and use in renewable energy hydrogen production systems for different scenarios.

Scenario	Initial hydrogen mass (kg/year)	Hydrogen production (kg/year)	Hydrogen use for FC (kg/year)	Hydrogen use for hydrogen load (kg/year)
G1	30	25,968	/	25,998
G2	25	28,301	2,336	25,990
G3	40	26,033	/	26,073
G4	50	28,401	2,429	26,022
G5	30	25,973	/	26,003
G6	30	27,783	1,808	25,993
S2	25	27,448	1,351	26,122
S4	50	26,755	737	26,068
S6	30	26,465	329	26,166

The capacity shortfalls of renewable HPSs for different scenarios are shown in Table 6. Due to the presence of the grid, only a shortage of hydrogen capacity is possible in a grid-connected system. In off-grid type systems, both electric and hydrogen capacity shortages may exist. From the table, it can be seen that for the grid-connected hydrogen production system, the hydrogen capacity shortages of the systems are closer and the shortages are all larger. In comparison, Scenarios G3 and G4 have smaller hydrogen capacity shortfalls. For the off-grid hydrogen production system, Scheme S6 has the smallest electric/hydrogen capacity shortfall in both cases. The main reason is that Scenario S6 utilizes the complementary characteristics of solar and wind energy in time, which has less impact on indirectness and volatility and enhances the stability of the system. Therefore, from the point of view of system capacity shortage rate, WT (G3) and WT/FC (G4) hydrogen production systems are technically better for grid-connected hydrogen production systems, and PV/WT/FC system (S6) hydrogen production systems are technically better for off-grid hydrogen production systems.

Table 6. Capacity shortfalls of renewable hydrogen production systems for different scenarios.

Scenario	Electricity capacity shortfall rate (%)	Hydrogen capacity shortfall rate (%)	Scenario	Electricity capacity shortfall rate (%)	Hydrogen capacity shortfall rate (%)
G1	/	5.0	G6	/	5.0
G2	/	5.0	S2	0.21	4.58
G3	/	4.75	S4	0.22	4.78
G4	/	4.94	S6	0.15	4.42
G5	/	5.0			

For off-grid systems, after all loads are satisfied by the power generated by the system components, if there is any remaining power it will be discarded. For the three off-grid hydrogen production systems (S2, S4, S6), the excess power for the different scenarios was 200,892 kWh/year, 1,369,712 kWh/year, and 821,721 kWh/year, with excess power rates of 13.1%, 73.2%, and 38.6%, respectively. The PV/FC (S2) and PV/WT/FC (S6) hydrogen production systems are analyzed from the point of view of the system excess power rate, which is small and technically better.

4.3 System Selection

Based on the above simulation results, different schemes of renewable energy HPS have their own advantages in terms of economy and technology, and through the analysis of economy and technology, the optimal grid-connected and off-grid renewable energy hydrogen production system schemes when considering different evaluation indicators are shown in Table 7. From the economic point of view, Scenarios G1 and G5 are the best solutions in the grid-connected system, and Scenarios S2 and S6 in the off-grid system have smaller costs for each of them, especially Scenario S6 is more advantageous in terms of energy costs. From a technical point of view, for the grid-connected system, schemes G4 and G6 are the best ones when considering the power generation. When considering the amount of hydrogen produced and the amount of hydrogen used, the actual amount of hydrogen used by the users of Scenarios G3 and G5 is larger and technically better. Scenarios G3, G4 are optimal when considering the hydrogen capacity shortage rate of the system. For the off-grid type hydrogen production system, schemes S4 and S6 are the best scenarios when considering the power generation of the system. Scenarios S2 and S6 are technically better when considering the amount of hydrogen production, the amount of hydrogen used, the capacity shortage rate and the excess power rate.

Usually, for grid-connected renewable energy hydrogen production systems, economic considerations dominate and the main consideration is the energy cost of the system. For off-grid renewable hydrogen production systems, economics and technology have equal status, and the main considerations are the capacity shortfall of the system and the energy cost. Therefore, for the grid-connected renewable energy hydrogen production system, the optimal solution is the PV/WT hydrogen production system (G5); for the off-grid renewable energy hydrogen production system, the PV/WT/FC hydrogen production system (S6).

Table 7. Optimal renewable energy hydrogen production system scenario for different performance metrics.

Performance indicators	Optimal grid-connected system scenario	Optimal off-grid system scenario
Economic indicators:		
Investment costs and operating costs	G1, G5	S2, S6
Net present value	G1, G5	S2, S6
Energy cost	G1, G5	S2, S6
Technical indicators:		
Generation capacity	G4, G6	S4, S6
Hydrogen production and use	G3, G5	S2, S6
Capacity shortfall rate	G3, G4	S2, S6
Excess power rate	/	S2, S6

5. Conclusion

Based on the basic mathematical models of the system components, this paper designs 12 different renewable energy HPS scenarios, including six grid-connected and six off-grid systems, using a location in China as the study site. Subsequently, the HPSs of the different scenarios are compared and analysed from both economic and technical perspectives in conjunction with the weather data and load data of the location in order to select the optimal grid-connected and off-grid HPS.

When considering the economic indicators such as investment cost, operating cost, NPV, and energy cost of renewable energy HPSs, the results show that the system containing PV panels has the lower of the four economic indicators for both grid-connected and off-grid renewable energy hydrogen production systems. Meanwhile, for a grid-connected hydrogen production system, other components being equal, a hydrogen production system without a FC has a smaller investment cost, operating cost and NPV than a hydrogen production system with a fuel cell. When considering the technical indicators such as power generation, hydrogen production, hydrogen use, capacity shortage rate and excess power rate of the renewable energy hydrogen production system, the actual hydrogen use of hydrogen loads is higher in Scenarios G3 and G5 of the grid-connected system. The actual amount of hydrogen used by the hydrogen loads in the hydrogen production systems of Scenarios S2 and S6 in the off-grid system is larger, while the electric capacity shortage rate, hydrogen capacity shortage rate and excess power rate are smaller.

In practice, for the grid-connected HPS, the economy is dominant and the main consideration is the energy cost of the system, and the optimal grid-connected hydrogen production system scenario is the PV/WT hydrogen production system with an energy cost of

0.152$/kWh. For the off-grid hydrogen production system, the economic and technical status are comparable, and the main considerations are the capacity shortage rate of the system and the energy cost, and the optimal off-grid hydrogen production system solution is the PV/WT/FC hydrogen production system, with an energy cost of $0.162/kWh and an electricity/hydrogen capacity shortage of 0.15% and 4.42%, respectively.

Acknowledgments

This research was funded by the Shandong Electric Power Engineering Consulting Institute Corp. Ltd., "Study on hydrogen dispersion, fire thermal radiation, and overall layout optimization for new skid-mounted hydrogen production systems and hydrogen refueling stations" (No. 042300007948).

Reference

[1] Skea J, Diemen R V, Portugal-Pereira J and Khourdajie A A. 2021 Outlooks Explorations and Normative Scenarios: Approaches to Global Energy Futures Compared. *Technological Forecasting and Soc Change* **168** 120736.

[2] Di Micco S, Romano F, Jannelli E, Perna A and Minutillo M. 2023 Techno-economic analysis of a multi-energy system for the co-production of green hydrogen, renewable electricity and heat. *Int J Hydrogen Energy* **48(81)** 31457-31467.

[3] Schnuelle C, Wassermann T, Fuhrlaender D and Zondervan E. 2020 Dynamic Hydrogen Production from PV & Wind Direct Electricity Supply-Modeling and Techno-economic Assessment. *Int J Hydrogen Energy* **45(55)** 29938-29952.

[4] Rezaei M, Mostafaeipour A, Qolipour M and Momeni M. 2019 Energy Supply for Water Electrolysis Systems Using Wind and Solar Energy to Produce Hydrogen: A Case Study of Iran. *Front Energy* **13(3)** 539-550.

[5] Marocco P, Ferrero D, Gandiglio M, Ortiz M M, Sundseth K, Lanzini A and Santarelli M. 2020 A Study of the Techno-economic Feasibility of H2-Based Energy Storage Systems in Remote Areas. *Energy Conversion and Management* **211** 112768.

[6] He H J, Huang Y Y, Nakadomari A, Masrur H, Krishnan N, Hemeida A M, Mikhaylov A and Senjyu T. 2023 Potential and economic viability of green hydrogen production from seawater electrolysis using renewable energy in remote Japanese islands. *Renewable Energy* **202 (2023)** 1436–1447.

[7] Karacavus B, Aydn K. 2020 Hydrogen Production and Storage Analysis of a System by Using TRNSYS. *Int J Hydrogen Energy* **45** 34608-34619.

[8] Fan X C, Wang W Q, Shi R J and Cheng Z J. 2017 Hybrid Pluripotent Coupling System with Wind and Photovoltaic-hydrogen Energy Storage and the Coal Chemical Industry in Hami, Xinjiang. *Renewable and Sustainable Energy Reviews* **72** 950-960.

[9] Rasheed M, Shihab S.2019 Parameters Estimation for Mathematical Model of Solar Cell. *Electronics Science Technology and Application* **6(1)** 20-28.

[10] Gokcek M, Kale C. 2018 Techno-Economical Evaluation of a Hydrogen Refuelling Station Powered by Wind-PV Hybrid Power System: A Case Study for İzmir-çeşme. *Int J Hydrogen Energy* **43(23)** 10615-10625.

Modeling of jet spreading and flame hazard distances for high pressure hydrogen releases

Fumin Guo[1], Chunyan Song[1], Qingxin Ba[2*], Jiaxin Zhang[3], Zhonglong He[3] and Xuefang Li[3]

[1]Shandong Electric Power Engineering Consulting Institute Corp. Ltd. Jinan 250061, China
[2]School of Mechanical Engineering, Shandong University, Jinan 250061, China
[3]Institute of Thermal Science and Technology (Institute for Advanced Technology), Shandong University, Jinan 250061, China

qingxin90113@126.com

Abstract: Hydrogen is a promising clean and sustainable energy carrier. Hydrogen is usually stored at high pressure due to its low density. Hydrogen releases from high pressure storage can result in underexpanded jets that will mix with air, forming a large combustible cloud. If the jets are ignited, jet flames will form with a large hazard area. This study simulated hydrogen jets and jet flames for storage pressures of 1~70 MPa and nozzle diameters of 1~2 mm using the HyRAM software. The results show that the jet diffusion distance increased with the hydrogen stagnation pressure and the nozzle diameter. The diffusion distances were correlated with the nozzle diameter and the hydrogen stagnation pressure. The jet flames and thermal radiations were then analyzed to show that the flame length and hazard distance both increased with the stagnation pressure and nozzle diameter. Correlations for radiation major harm distance, harm distance and no harm distance were developed to predict the jet flame hazard range. The present study can provide references for the safety design of hydrogen applications.
Keywords: High pressure hydrogen, Jet, Jet flame, Hazard distance

1. Introduction

The fossil fuel crisis and air pollution have become increasingly serious in recent years. Hydrogen has been increasingly used as a promising clean and sustainable energy carrier [1-3]. Over 30 governments have pledged to incorporate hydrogen as a clean energy vector into their energy systems. At present, the storage technology of high-pressure gaseous hydrogen is well developed, with the advantages of well-established technology, low energy consumption, and fast hydrogen filling. Hydrogen is flammable and explosive, with wide flammability limits in the air. Therefore, hydrogen safety is a serious issue requiring utmost attention in hydrogen applications. Hydrogen releases and ignitions are potential accident scenarios, which are important for the design of hydrogen storage scenarios and developing the safety codes and standards.

The flow is choked at the nozzle when hydrogen releases from a high pressure source, forming an underexpanded, supersonic jet. Many researchers have measured and simulated underexpanded hydrogen jets to study the shock structures and jet parameter distributions [4-11]. There are complex shock stuctuures near the outlet of underexpanded hydrogen jets. The Mach disk location has been correlated as a function of the pressure ratio and the nozzle diameter. The field parameter distributions are self-similar. The jet axial concentration and velocity decay hyperbolically, while the radial concentration and velocity distributions conform to the Gaussian distributions. Hydrogen releases from high-pressure sources will form jet flames if ignited, leading to disastrous consequences. The jet flame characteristics have been measured in several studies [12-14]. The flame shape was visualized for various release conditions. The flame length and width were correlated as functions of the stagnation pressure and nozzle diameter [12, 13]. There are also some studies simulated hydrogen jet flames and predicted the flame temperature distributions using Computational Fluid Dynamics (CFD) methods [15-18]. In addition, the effect distances of hydrogen jets and jet flames are important for the safety design of hydrogen plants and stations. However, the research on the effect distances of jet spreading and jet flames is still inadequate.

HyRAM is an integrated platform for hydrogen safety research, which was developed by the Sandia National Laboratory with the support of the US Department of Energy [19]. In HyRAM the system configurations can be defined by the user and then the multiple risk and harm metrics can be calculated [20]. The HyRAM calculating results will provide insights for stakeholders in the safety, codes, and standards community. The quantitative risk assessment in HyRAM incorporates generic failure probabilities for hydrogen components, the thermal hazard probabilistic model and the overpressure hazard probabilistic model. LaFleur et al. [21] calculated risk values for developing risk-equivalent designs and documented the performance-based process by HyRAM. Xing et al. [22] used HyRAM to analyze the high-risk unit for urban hydrogen refueling stations and determine the fatality zone, flammable area and safe distances. Park et al. [23] studied accidents at hydrogen refueling stations and analyzed individual and societal risks using the HyRAM software. HyRAM also incorporates experimentally validated models of releases and flames. Simulations using HyRAM enable fast calculation of jet and flame characteristics compared to CFD methods.

In this study, the jet and jet flame after leakage of hydrogen stored at high pressure were simulated using HyRAM. The hydrogen mole fractions were predicted for stagnation pressures of 1~70 MPa and nozzle diameters of 1~2 mm. The jet diffusion distance was correlated with the stagnation pressure and nozzle diameter. The jet flame temperatures and radiative heat flux distributions were then calculated for stagnation pressures of 1~70 MPa and nozzle diameters of 1~2 mm. The major harm, harm and no harm distances of flame thermal radiation were also analyzed and correlated.

2. Simulation methods

2.1. *Problem description*

The high pressure hydrogen was released horizontally from circular nozzles with diameters of 1 and 2 mm at stagnation pressures of 1 to 70 MPa. Underexpanded jets will form if the hydrogen is not ignited after leakage, while a jet flame will form if the hydrogen is ignited during leakage. The jets and jet flames were simulated using HyRAM in this paper. The hydrogen stagnation temperature and ambient air temperature were both 300 K. The stagnation pressures and the nozzle diameters for simulated cases are listed in Table 1.

Table 1. Stagnation pressures and the nozzle diameters for simulated cases.

Cases	1	2	3	4	5	6	7	8	9	10	11	12
p (MPa)	1	1	5	5	10	10	30	30	50	50	70	70
d (mm)	1	2	1	2	1	2	1	2	1	2	1	2

2.2. *HyRAM simulation for jets*

The one-dimensional model [24] was used in HyRAM for jet modeling. The accuracy of the jet model in HyRAM has been validated [25]. The jet bending in axial direction caused by buoyancy effects was considered in the model. The disributions of the jet velocity (v), density (ρ), and product of density and hydrogen mass fraction (Y) are assumed to be Gaussian [20], as follows:

$$v = v_{cl} \exp\left(-\frac{r^2}{B^2}\right) \quad (1)$$

$$\rho = \left(\rho_{cl} - \rho_{amb}\right)\exp\left(-\frac{r^2}{\lambda^2 B^2}\right) + \rho_{amb} \quad (2)$$

$$\rho Y = \rho_{cl} Y_{cl} \exp\left(-\frac{r^2}{\lambda^2 B^2}\right) \quad (3)$$

The derivatives of the spatial dimensions are

$$\frac{dx}{dS} = \cos\theta \quad (4)$$

$$\frac{dy}{dS} = \sin\theta \quad (5)$$

The conservation equations can be written as [20]

$$\frac{d}{dS}\int_0^\infty (\rho v) 2\pi r dr = \rho_{amb} E \quad (6)$$

$$\frac{d}{dS}\int_0^\infty (\rho v Y) 2\pi r dr = 0 \quad (7)$$

$$\frac{d}{dS}\int_0^\infty \left(\rho v^2 \cos\theta\right) 2\pi r dr = 0 \quad (8)$$

$$\frac{d}{dS}\int_0^\infty \left(\rho v^2 \sin\theta\right) r dr = \int_0^\infty \left(\rho_{amb} - \rho\right) g r dr \quad (9)$$

$$\frac{d}{dS}\int_0^\infty \rho v \left(h + \frac{v^2}{2} - h_{amb}\right) 2\pi r dr = 0 \quad (10)$$

where r is perpendicular to the jet flow direction, B is the half-width, λ is the ratio of density spreading to velocity, E is the entrainment, θ is the angle relative to horizontal, h is the enthalpy, the subscript *cl* represents centerline, *amb* represents ambient. The ideal gas equation of state is used to calculate the physical properties of atmospheric pressure mixtures.

2.3. HyRAM simulation for jet flames

The jet flame radiative heat flux was predicted by a weighted multi-source model. The heat flux can be described as [20]

$$q = \tau S_{rad} \frac{V_F}{A_f} \quad (11)$$

where τ is the transmissivity, V_F is the view factor, which is proportional to the heat flux, A_f is the flame surface area, and S_{rad} is the total emitted radiative power from a flame, calculated by

$$S_{rad} = X_{rad} \dot{m}_{fuel} \Delta H_c \quad (12)$$

$$X_{rad} = 9.45 \times 10^{-9} \left(\tau_f a_p T_{ad}^4\right)^{0.47} \quad (13)$$

$$\tau = \frac{\rho_f W_f^2 L_{vis} f_s}{3\rho_j d_j^2 u_j} \quad (14)$$

$$\rho_f = \frac{p_{amb} W_{mix}}{R T_{ad}} \tag{15}$$

where $\dot{m}_{fuel}$ represents the hydrogen mass flow rate, ΔH_c represents the combustion heat, which equals to 120 MJ/kg, X_{rad} represents the radiant fraction, a_p equals to 0.23, which represents the Planck-mean absorption coefficient, τ_f represents the residence time of flame, ρ_f represents the flame density, P_{amb} represents the ambient pressure, W_{mix} represents the mean molecular weight of the combustion stoichiometric products, and R represents the gas constant. The accuracy of the flame model in HyRAM has been validated [25].

3. Results and discussion

3.1. High pressure hydrogen jets

3.1.1. Concentration distributions. The hydrogen mole fractions were simulated using the physical module in HyRAM. Figure 1 shows the hydrogen mole fraction distributions for various stagnation pressures and nozzle diameters. The white curve in Figure 1 indicates the hydrogen mole fraction of 0.04, which is the lower flammable limit (LFL) of hydrogen in air. The area within the white curve represents the area of the flammable cloud. The flammable cloud spreading range increases with the nozzle diameter for a given pressure and increases with the stagnation pressure for a given nozzle diameter.

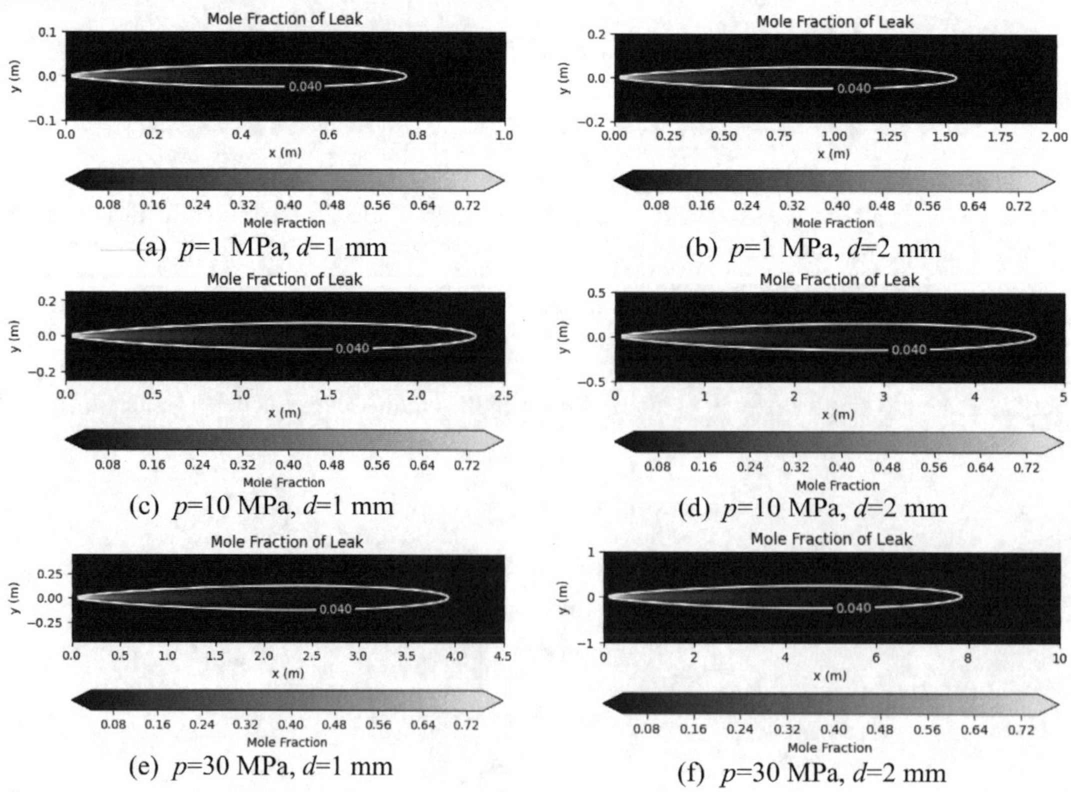

(a) p=1 MPa, d=1 mm (b) p=1 MPa, d=2 mm
(c) p=10 MPa, d=1 mm (d) p=10 MPa, d=2 mm
(e) p=30 MPa, d=1 mm (f) p=30 MPa, d=2 mm

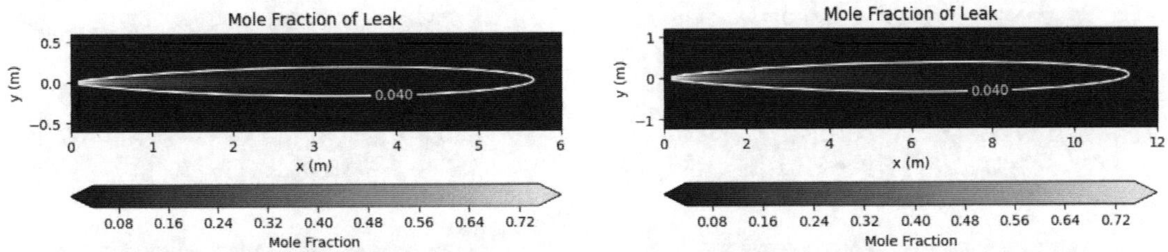

(g) p=70 MPa, d=1 mm (h) p=70 MPa, d=2 mm

Figure 1. Hydrogen mole fraction distributions for various stagnation pressures and nozzle diameters.

3.1.2. Flammable cloud spreading distances. The flammable cloud spreading distance variation as a function of the stagnation pressure is shown in Figure 2. The spreading distance increases with the hydrogen stagnation pressure for a given nozzle diameter. For the nozzle diameter of 1 mm, the spreading distance is 0.77 m with a stagnation pressure of 1 MPa and increases to 5.68 m with a stagnation pressure of 70 MPa. In addition, the spreading distance increases with increasing nozzle diameter for a given stagnation pressure. For the stagnation pressure of 70 MPa, the spreading distance is 5.68 m with 1 mm diameter nozzle and increases to 11.25 m with 2 mm diameter nozzle.

The spreading distance was non-dimensionalized by the nozzle diameter to develop an empirical correlation. Figure 3 shows the correlation between the dimensionless spreading distance, D_{LFL}/d, and the hydrogen stagnation pressure. The black line in Figure 3 is the fitting curve using the least square method. The correlation between the spreading distance, the nozzle diameter and the hydrogen stagnation pressure was developed as

$$D_{LFL}/d = 0.81p^{0.46} \tag{16}$$

where the coefficients "0.81" and "0.46" were determined by the least square method. The determination coefficient (R^2) is 0.9995, which shows that the fitting curve agrees well with the data.

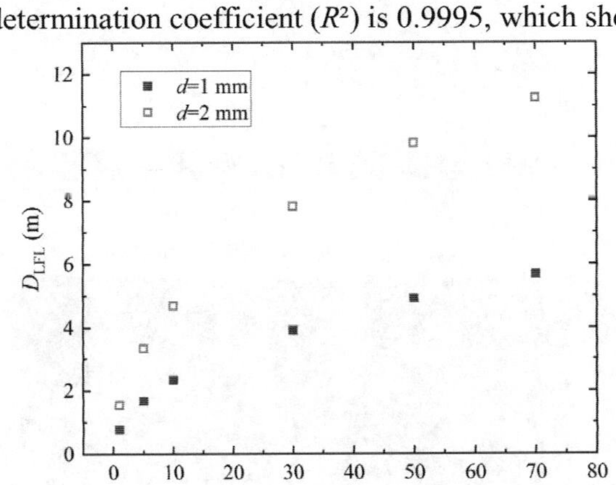

Figure 2. Spreading distances of the flammable cloud.

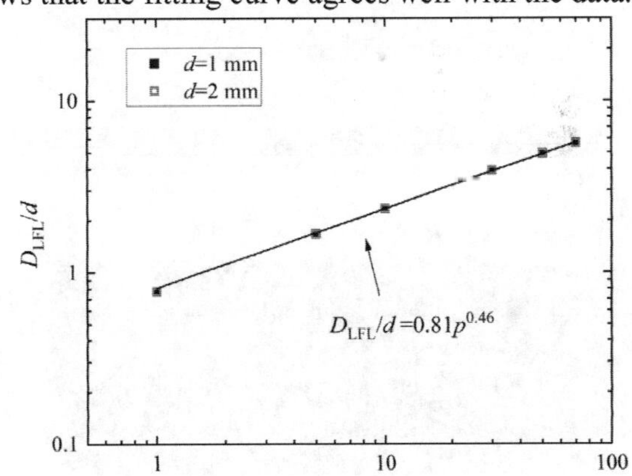

Figure 3. Correlation of the dimensionless spreading distance.

3.2. High pressure hydrogen jet flames

3.2.1. Jet flames. The jet flame trajectories and temperatures were simulated for various stagnation pressures and nozzle diameters, as shown in Figure 4. The results show that larger leak diameters and stagnation pressures result in longer flame lengths. For the nozzle diameter of 1 mm, the flame length increases from 0.42 to 3.07 m with the stagnation pressure increasing from 1 to 70 MPa. For the stagnation pressure of 70 MPa, the flame length increases from 3.07 to 6.15 m with the nozzle diameter increasing from 1 to 2 mm.

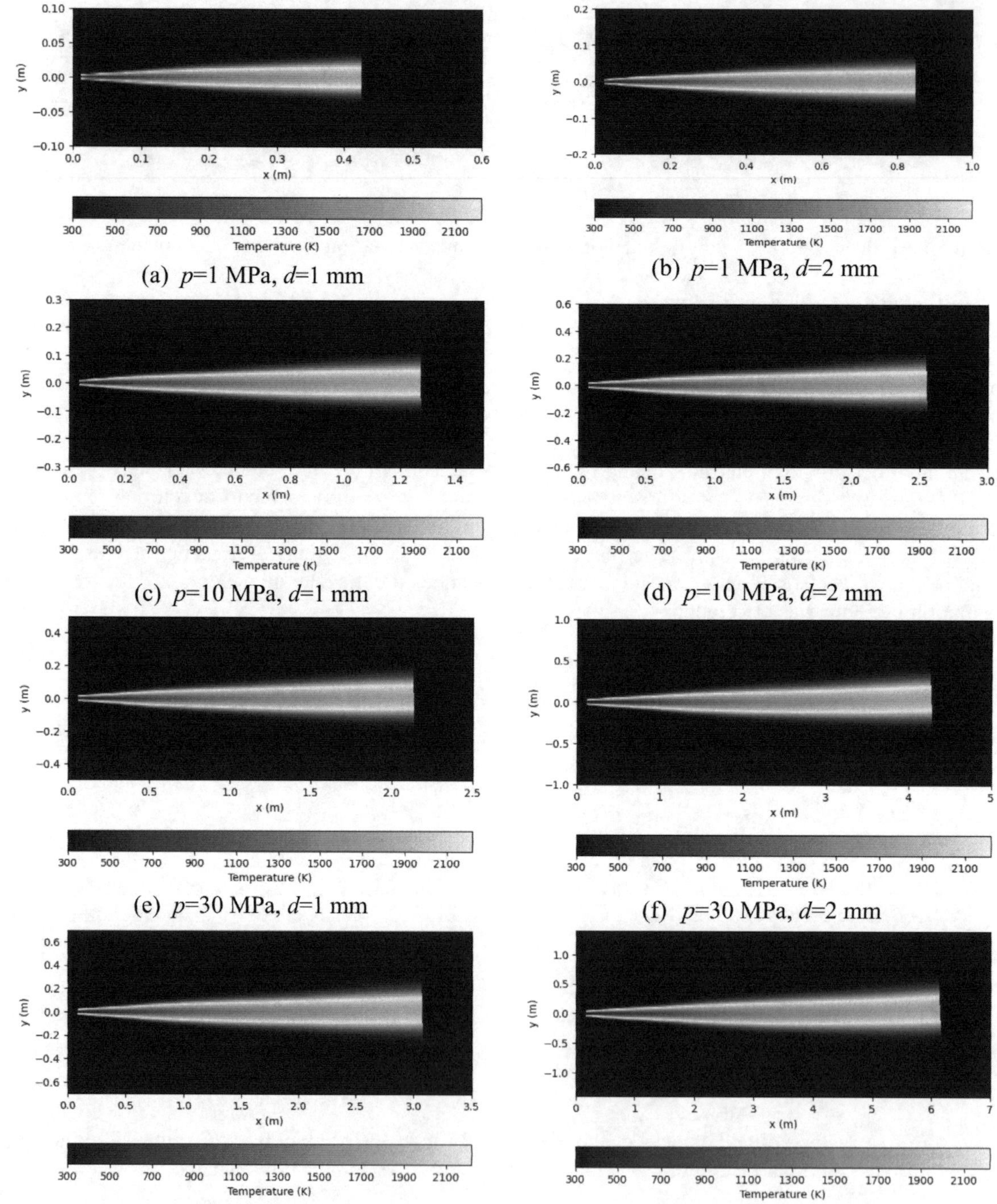

(a) p=1 MPa, d=1 mm (b) p=1 MPa, d=2 mm
(c) p=10 MPa, d=1 mm (d) p=10 MPa, d=2 mm
(e) p=30 MPa, d=1 mm (f) p=30 MPa, d=2 mm
(g) p=70 MPa, d=1 mm (h) p=70 MPa, d=2 mm

Figure 4. Hydrogen jet flame temperatures for various stagnation pressures and nozzle diameters.

3.2.2. Flame thermal radiation effects. The thermal radiation flux can be evaluated relative to the harm degree and safety distance of jet flame. The thermal radiation heat flux distributions of hydrogen jet flame were simulated for various stagnation pressures and nozzle diameters, as shown in Figure 5. The three colors in Figure 5 represent different harm degrees for equipment and people. The heat flux of 25.24 kW/m^2 is considered as the major harm criterion, which can cause equipment damage. The heat flux of 4.73 kW/m^2 is considered as the harm criterion, which can cause injury to people. The heat flux of 1.58 kW/m^2 is considered as the no harm criterion, which has no harm to equipment and people [26].

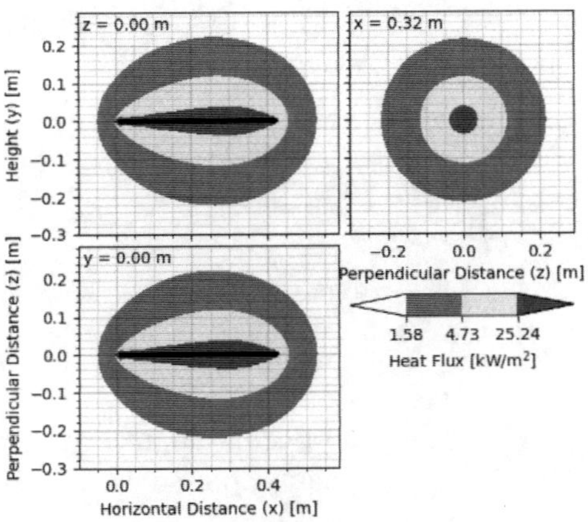

(a) p=1 MPa, d=1 mm

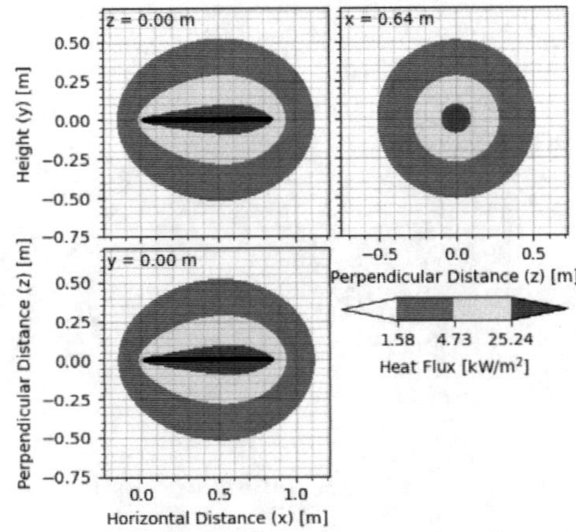

(b) p=1 MPa, d=2 mm

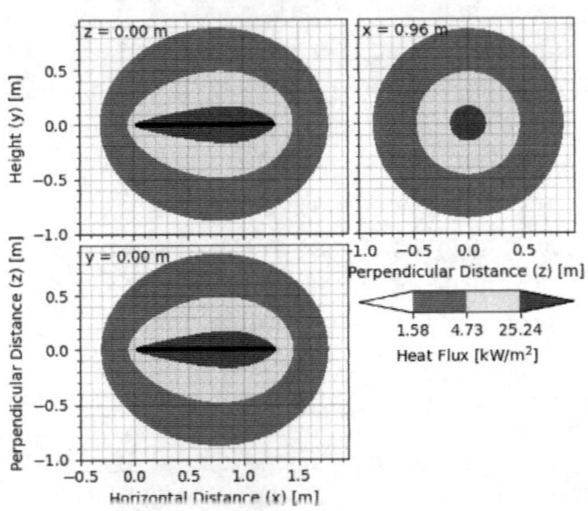

(c) p=10 MPa, d=1 mm

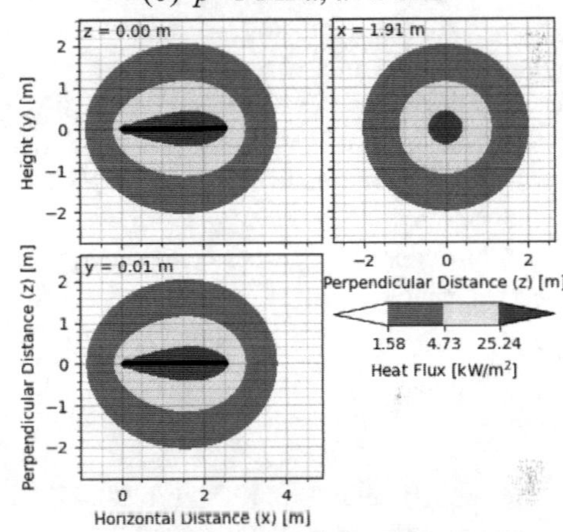

(d) p=10 MPa, d=2 mm

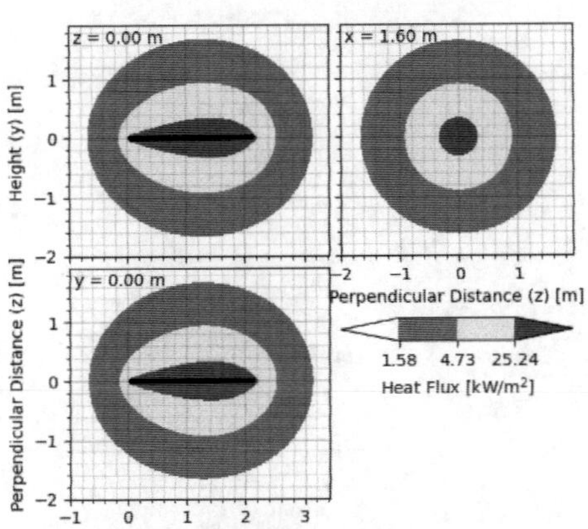

(e) p=30 MPa, d=1 mm

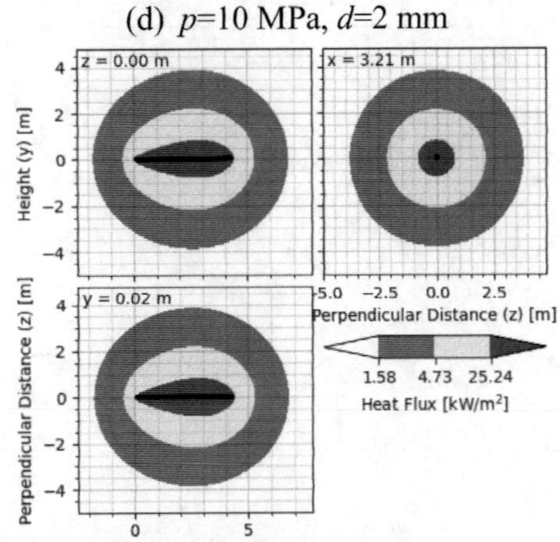

(f) p=30 MPa, d=2 mm

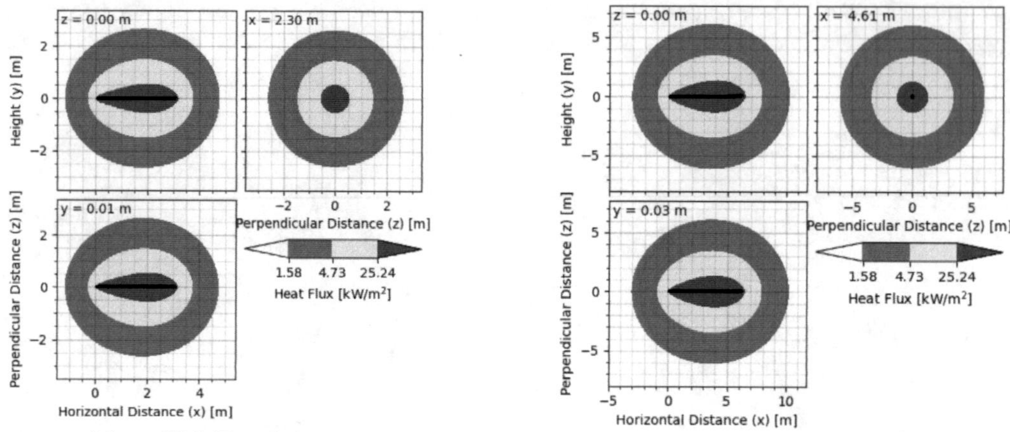

(g) p=70 MPa, d=1 mm (h) p=70 MPa, d=2 mm

Figure 5. Heat flux distributions of hydrogen jet flame for various stagnation pressures and nozzle diameters.

3.2.3. Thermal radiation hazard distances. The major harm distance variation as a function of the hydrogen stagnation pressure is shown in Figure 6. For the nozzle diameter of 1 mm, the major harm distance increases from 0.42 to 3.16 m with the stagnation pressure increasing from 1 to 70 MPa. For the nozzle diameter of 2 mm, the major harm distance increases from 0.85 to 6.48 m with the stagnation pressure increasing from 1 to 70 MPa.

The major harm distance was nondimensionalized by the nozzle diameter. Figure 7 shows the correlation between the dimensionless major harm distance, $D_{\text{Major harm}}/d$, and the hydrogen stagnation pressure. The black line in Figure 7 is the fitting curve using the least square method. The correlation between the major harm distance, the nozzle diameter and the hydrogen stagnation pressure was developed as

$$D_{Major\ harm}/d = 0.43p^{0.47} \tag{17}$$

where the coefficients "0.43" and "0.47" were determined by the least square method. The determination coefficient (R^2) is 0.9993, which shows that the fitting curve agrees well with the data.

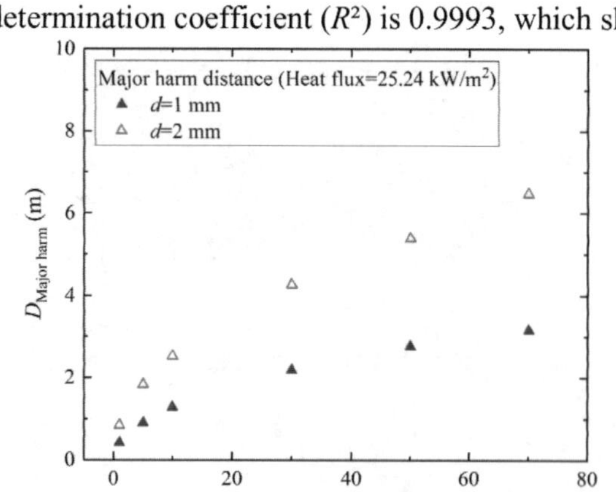

Figure 6. Major harm distances of jet flames.

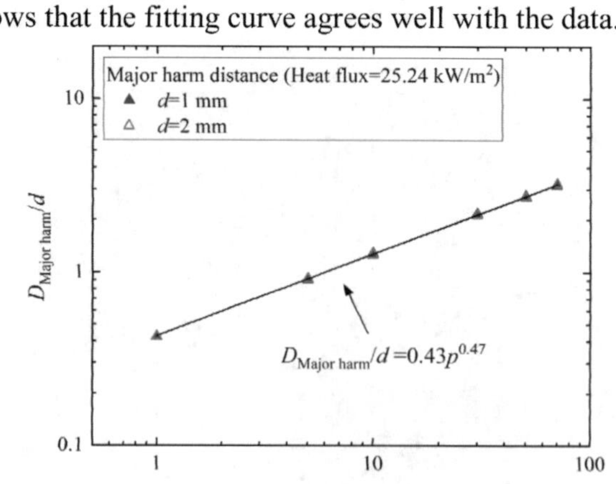

Figure 7. Correlation of the dimensionless major harm distance.

The harm distances for various stagnation pressures and nozzle diameters are shown in Figure 8. For the nozzle diameter of 1 mm, the harm distance increases from 0.45 to 3.77 m with the stagnation pressure increasing from 1 to 70 MPa. For the nozzle diameter of 2 mm, the major harm distance increases from 0.93 to 7.95 m with the stagnation pressure increasing from 1 to 70 MPa.

Figure 9 shows the correlation between the dimensionless harm distance, D_{Harm}/d, and the hydrogen stagnation pressure. The black line in Figure 9 is the fitting curve using the least square method. The expression between the harm distance, the nozzle diameter and the hydrogen stagnation pressure was developed as

$$D_{Harm}/d = 0.46p^{0.50} \tag{18}$$

where the coefficients "0.46" and "0.50" were determined by the least square method. The determination coefficient (R^2) is 0.998, which shows that the fitting curve agrees well with the data.

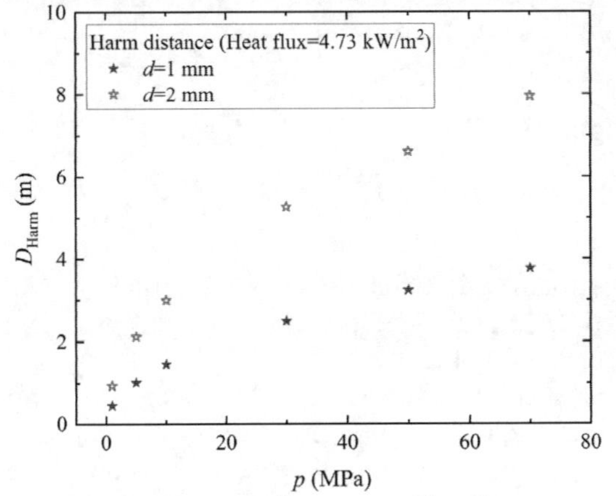

Figure 8. Harm distances of jet flames.

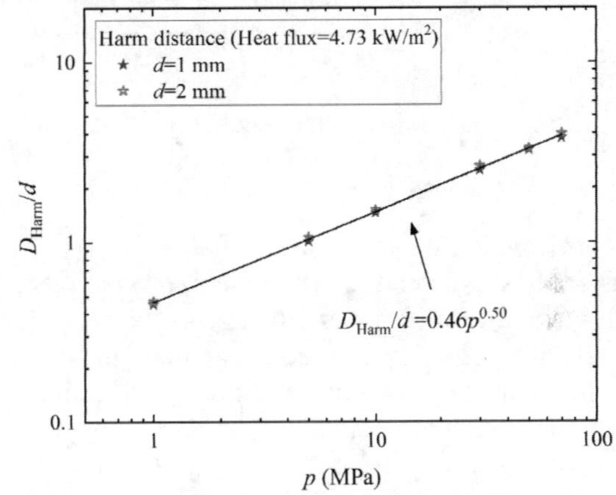

Figure 9. Correlation of the dimensionless harm distance.

Figure 10 shows the no harm distances for various stagnation pressures and nozzle diameters. For the nozzle diameter of 1 mm, the no harm distance increases from 0.53 to 4.72 m with the stagnation pressure increasing from 1 to 70 MPa. For the nozzle diameter of 2 mm, the major harm distance increases from 1.11 to 10.24 m with the stagnation pressure increasing from 1 to 70 MPa.

Figure 11 shows the correlation between the dimensionless harm distance, $D_{No\ Harm}/d$, and the hydrogen stagnation pressure. The black line in Figure 11 is the fitting curve using the least square method. The expression between the no-harm distance, the nozzle diameter and the hydrogen stagnation pressure was developed as

$$D_{No\ harm}/d = 0.56p^{0.51} \tag{19}$$

where the coefficients "0.56" and "0.51" were determined by the least square method. The determination coefficient (R^2) is 0.995, which shows that the fitting curve agrees well with the data.

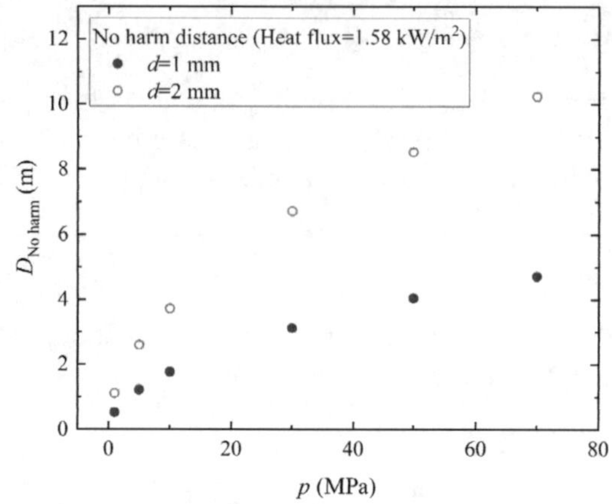

Figure 10. No harm distances of jet flames.

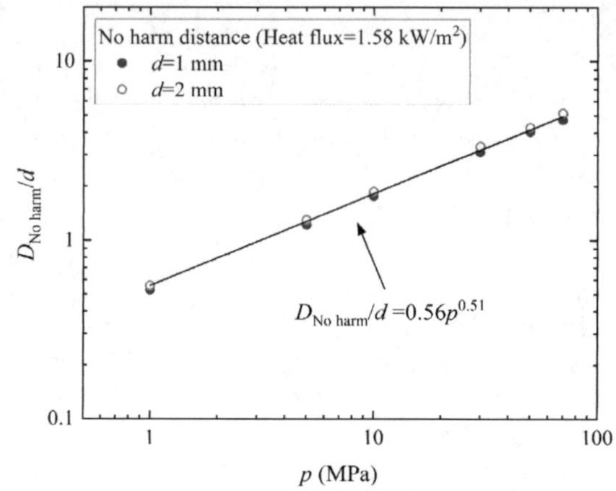

Figure 11. Correlation of the dimensionless no harm distance.

4. Conclusions

This study simulated high pressure hydrogen jets and jet flames for stagnation pressures of 1~70 MPa with 1 and 2 mm diameter circular nozzles using the HyRAM software. The experimentally validated models in HyRAM were used to calculate the hydrogen jet and flames.

The hydrogen jet concentration distributions were predicted for various hydrogen stagnation pressures and nozzle diameters. The flammable cloud spreading distance increases with increasing stagnation pressure and increasing nozzle diameter. The spreading distance correlation was developed as a power function of the nozzle diameter and the hydrogen stagnation pressure to calculate the jet effect distance of hydrogen stored at high pressure. The hydrogen jet flame temperatures and radiation heat flux were also predicted for various conditions. The thermal radiation harm degrees and distances were then analyzed. The thermal radiation hazard degrees were divided into major harm, harm and no harm. The major harm distance, harm distance and no harm distance were correlated as power functions of the nozzle diameter and the stagnation pressure to assess the effect distances of high pressure hydrogen jet flames.

The results can provide a scientific basis for the safety design of hydrogen plants and stations.

Acknowledgments

This research was funded by the Shandong Electric Power Engineering Consulting Institute Corp. Ltd., “Study on hydrogen dispersion, fire thermal radiation, and overall layout optimization for new skid-mounted hydrogen production systems and hydrogen refueling stations” (No. 042300007948).

References

[1] Barbir F 2009 Transition to Renewable Energy Systems with Hydrogen as an Energy Carrier. *Energy* **34**(3) 308-12.

[2] Apak S, Atay E and Tuncer G 2017 Renewable Hydrogen Energy and Energy Efficiency in Turkey in the 21st Century. *Int. J. Hydrog. Energy* **42**(4) 2446-52.

[3] Panwar N L, Kaushik S C and Kothari S 2011 Role of Renewable Energy Sources in Environmental Protection: A Review. *Renew. Sustain. Energy. Rev* **15**(3) 1513-24.

[4] Li X, Christopher D M, Hecht E S and Ekoto I W 2017 Comparison of two-layer model for hydrogen and helium jets with notional nozzle model predictions and experimental data for pressures up to 35 MPa. *Int. J. Hydrog. Energy* **42**(10) 7457-66.

[5] Hu J, Christopher D M and Li X 2018 Simplified partitioning model to simulate high pressure under-expanded jet flows impinging vertical obstacles. *Int. J. Hydrog. Energy* **43**(29) 13649-58.

[6] Li X, Yao C, Egbert S C, He Q, Zhao Z, Christopher D M and Hecht E S 2023 Self-similar

characteristics of underexpanded, cryogenic hydrogen and methane jets. *Int. J. Hydrog. Energy* **48**(10) 4104-17.

[7] Franquet E, Perrier V, Gibout S and Bruel P 2015 Free underexpanded jets in a quiescent medium: a review. *Prog. Aero. Sci* **77** 25-53.

[8] Han S H, Chang D and Kim J S 2014 Experimental investigation of highly pressurized hydrogen release through a small hole. *Int. J. Hydrog. Energy* **39**(17) 9552-61.

[9] Kotchourko N, Kuznetsov M, Kotchourko A, Grune J, Lelyakin A and Jordan T 2014 Concentration measurements in around hydrogen jet using Background Oriented Schlieren (BOS) technique. *Int. J. Hydrog. Energy* **39**(11) 6201-09.

[10] Ba Q, He Q, Zhou B, Chen M, Li X, Cheng L 2020 Modeling of cryogenic hydrogen releases. *Int. J. Hydrog. Energy* **45**(55) 31315-26.

[11] Li X, Chen Q, Chen M, He Q, Christopher D M, Cheng X, Chowdhury B R and Hecht E S 2019 Modeling of underexpanded hydrogen jets through square and rectangular slot nozzles. *Int. J. Hydrog. Energy* **44**(12) 6353-65.

[12] Schefer R, Houf W, Bourne B and Colton J 2006 Spatial and Radiative Properties of an Open-flame Hydrogen Plume. *Int. J. Hydrog. Energy* **31**(10) 1332-40.

[13] Mogi T and Horiguchi S 2009 Experimental Study on the Hazards of High-pressure Hydrogen Jet Diffusion Flames. *J. Loss. Prev. Process. Ind* **22**(1) 45-51.

[14] Panda P P and Hecht E S 2017 Ignition and Flame Characteristics of Cryogenic Hydrogen Releases. *Int. J. Hydrog. Energy* **42**(1) 775-85.

[15] Brennan S L, Makarov D V and Molkov V 2009 LES of High Pressure Hydrogen Jet Fire. *J. Loss. Prev. Process. Ind* **22**(3) 353-59.

[16] Makarov D V and Molkov V 2013 Plane Hydrogen Jets. *Int. J. Hydrog. Energy* 38(19) 8068-83.

[17] Ba Q, Xiao J, Jordan T, Huang T, Zhao M, Xiao G and Li X 2021 Measurement and Modeling on Hydrogen Jet and Combustion from a Pressurize Vessel, 2021 *International Conference on Hydrogen Safety*, Edinburgh, , September 21-23.

[18] Ba Q, Zhao Z, Zhang Y, Liu Y, Christopher D M, Ge P and Li X 2023 Modeling of cryogenic compressed hydrogen jet flames. *Int. J. Hydrog. Energy*, in press.

[19] Groth K M and Hecht E S 2017 HyRAM: A methodology and toolkit for quantitative risk assessment of hydrogen systems. *Int. J. Hydrog. Energy* **42**(11) 7485-93.

[20] Ehrhart B D and Hecht E S 2022 Hydrogen Plus Other Alternative Fuels Risk Assessment Models (HyRAM+) Version 5.0 Technical Reference Manual. SAND2022-16425, Sandia National Laboratories.

[21] Lafleur A C, Muna A B and Groth K M 2016 Application of quantitative risk assessment for performance-based permitting of hydrogen fueling stations. *Int. J. Hydrog. Energy* **42**(11) 7529-35.

[22] Xing Y, Wu J, Bai Y, Cai J and Zhu X 2022 All-process risk modelling of typical accidents in urban hydrogen refueling stations. *Process. Saf. Environ. Prot* **166** 414-29.

[23] Park B, Kim Y, Paik S, Kang C 2021 Numerical and Experimental Analysis of Jet Release and Jet Flame Length for Qualitative Risk Analysis at Hydrogen Refueling Station. *Process. Saf. Environ. Prot* **155** 145-54.

[24] Houf W G and Winters W S 2013 Simulation of high-pressure liquid hydrogen releases. *Int. J. Hydrog. Energy* **38**(19) 8092-99.

[25] Ehrhart B D, Hecht E S and Mohmand J A 2021 Validation and Comparison of HyRAM Physics Models. SAND2021-5811, Sandia National Laboratories.

[26] International Code Council, "International Fire Code," 2002, ISBN:892395-60-6.

Cold start of proton exchange membrane fuel cell using build-in catalytic heater

N A Ivanova[1], I E Baranov[1], A A Kalinnikov[1], R M Mensharapov[1], D D Spasov[1,2], M V Sinyakov[1,3], I I Nikolaev[1], S V Ostrovsky[1], S A Grigoriev[1,2,4] and V N Fateev[1]

[1] National Research Center "Kurchatov Institute", Russia, 123182 Moscow, Kurchatov Sq., 1
[2] National Research University "Moscow Power Engineering Institute", Russia, 111250 Moscow, Krasnokazarmennaya St., 14
[3] D. Mendeleev University of Chemical Technology of Russia, 125047 Moscow, Miusskaya Sq., 9
[4] HySA Infrastructure Center of Competence, Faculty of Engineering, North-West University, Private Bag X6001, Potchefstroom 2531, South Africa

sergey.grigoriev@outlook.com

Abstract. The paper report on the cold start of fuel cell with proton exchange membrane (PEMFC) at – 40 °C using a catalytic heating unit integrated directly into the PEMFC bipolar plates. This technical solution increases the heat transfer efficiency up to 60% due to direct contact of the membrane-electrode assembly with the heating unit, and ensure a successful cold start of the fuel cell from – 40 °C to an operating temperature of 35 °C within 6 minutes at air flow rate of 150 mL/min. The hydrogen flow rate is 45 cm^3/s, which corresponds to a hydrogen concentration in the air flow of *ca.* 1.8 vol.%, which is below the autoignition point and ensures the safety of the proposed method. Uniform distribution of heat over the bipolar plates surface prevents dehydration and thermal degradation of the membrane electrode assembly components and improve the PEMFC performance after cold start.

1. Introduction

The main advantages of fuel cells with proton exchange membrane (PEMFC) are high efficiency and zero emission, as well as fast load response, long service life, and a high level of fire and explosion safety. Regarding the PEMFC applications in cold climate conditions (*i.e.* at subzero temperature) there is a problem of water freezing in the volume of the device elements. Ice formation makes it difficult to successfully restart the system and causes permanent component damage and degradation of system performance [1–3]. To eliminate these negative effects, several cold start strategies and corresponding technical solutions are proposed. These scenarios include moisture removal by gas purging of water-containing elements (for example, using dry nitrogen) [3–16], vacuum drying, and the use of antifreezes [11] (for example, an aqueous solution of methanol), heat exchangers, bifunctional microporous layers, the use of materials that prevent freezing or do not degrade when frozen, running the system at constant current or maximum power, etc. These methods ensure the device self-starts at −5 … −10 °C. The use of methanol as an antifreeze (40% methanol-water solution) provides successfully PEMFC self-start at −15 °C. In the case of deeper freezing of devices at −30 … −50 °C the assisted heating is needed to provide fast start-up [17–37]. The main solutions are internal and external heating, namely hot air-blowing, electrical heating methods, and the use of catalytic hydrogen-oxygen recombiner. Compared with other assisted heating methods, the catalytic recombination reaction has the advantages of system design, implementation and high heating efficiency [25].

In [6, 36], an external heating method based on the idea of a catalytic recombiners to generate a large amount of heat in a short period of time was described. A separate device was used for a catalytic recombination reaction under pressure, which produced a heated gas that exchanged heat with the components of the PEMFC. The highly efficient catalytic recombiner based on the catalytic oxidation of hydrogen for supplying heat to a hydrogen storage system was proposed in [38]. The authors of the patent [39] have developed a catalytic hydrogen-oxygen recombiner for PEMFC start-up at subzero temperatures with possibility to regulate the final temperature. The main advantage of such systems is the high speed of heating. On the other hand, in the case of external heating, the heat transfer coefficient is still not high due to the use of gases as coolants to heat the device. This method needs for additional sources of hydrogen and increases the total cost and size of the installation. Internal heating is provided by adding hydrogen to the air flow (cathode) and/or adding oxygen to the hydrogen flow (anode) to generate heat in the catalytic layers due to recombination reaction on the Pt active cites [40, 41]. Modeling studies demonstrate [42] that this internal catalytic reaction affects the temperature distribution and improves cold start performance. The main disadvantages include high concentrations of gas additives in flows up to 20 vol.% and the possibility of local overheating, resulting in its dehydration and thermal degradation of the membrane electrode assembly (MEA) components.

In this paper we studied a cold start method assisted by using a catalytic heating unit mounted in direct contact with the MEAs of PEMFC stack. The design of heating unit consists of three reactors for hydrogen-oxygen recombination reaction (recombiners) directly built into bipolar plates. This solution increases the PEMFC heating efficiency due to the direct contact of the MEA (*i.e.* gas diffusion layers [43]) with heated bipolar plates. The parameters of the heating unit were investigated. Successful assisted cold start process of the PEMFC at –40 °C has been experimentally demonstrated.

2. Materials and methods

2.1. Catalytic recombiner

The catalytic hydrogen recombiner is a steel tube (Fig. 1a) with an inner diameter of 3 mm and a length of 60 mm, loaded with a catalyst. Spherical γ-alumina granules fraction size 0.6–1.0 mm with 1 wt % deposited platinum were used as a catalyst. The particle size of platinum does not exceed 5 nm. Three catalytic recombiners are built into the bipolar plates (BP) of the PEMFC, as shown in Fig. 1b, and together form a catalytic heating unit. Placing recombiners in bipolar plates is the same in the cost, and do not exceed the total costs of producing the PEMFC stack and a separate catalytic heating unit both in machining and materials.

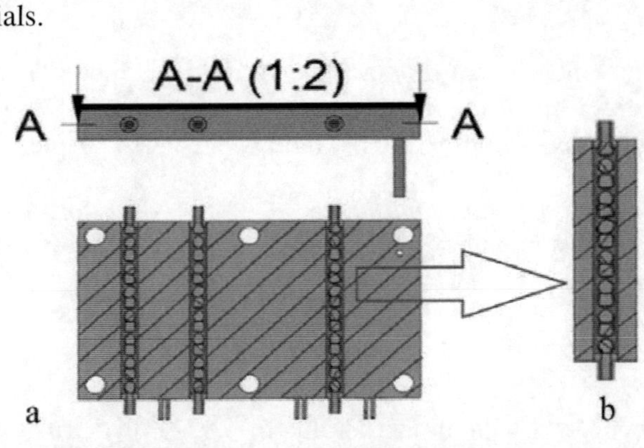

Figure 1. Catalytic heating unit (a) and catalytic recombiner (b) embedded directly into the BP.

The construction material of the presented BP is corrosion-resistant steel. The use of small diameter recombiners (inner diameter 3 mm) provided to avoid a significant increase in the BP thickness. The recombiner have come into direct contact with a conductive material for effective heat transfer, most often metal. However, the thickness of the metal plate can be reduced if the recombines are not located in the thickness of the plate but are welded on the outside. Porous plates can be also used, including those made using 3D printing.

2.2. Catalytic recombiner testing
The scheme of the experimental setup is shown in Fig. 2. The catalytic recombiner was placed inside the flow channel of the experimental chamber with a diameter of 25 mm, through which a hydrogen-air mixture was driven. The temperature on the recombiner surface was measured with a Uti165A thermal imager through a 2 × 65 mm^2 observation window as shown in Fig. 2. Thermal maps were obtained using a thermal imager at certain points in time, which were stored on an SD card, and then processed using the program included in the software of the Uti165A. Additionally, the temperature at the center of the catalytic recombiner was measured using a chromel-copel thermocouple immersed in the catalyst bed. Low subzero temperatures were also recorded with a thermocouple due to the limitations of the thermal imager. The thermocouple signal, after amplification on a precision operational amplifier OP177, was digitized and transferred to the processing program, which showed a graph of temperature changes in real time.

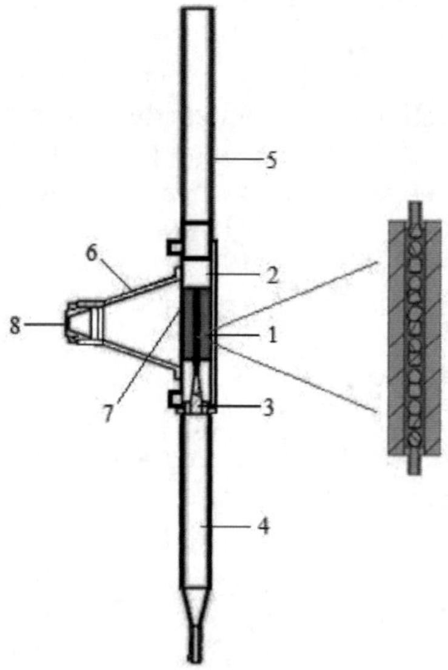

Figure 2. Scheme of an experimental setup for testing a catalytic recombiner: 1 - catalytic recombiner, 2 - flow chamber, 3 - holder for the recombiner, 4 - gas mixing chamber, 5 - removable outlet pipe, 6 - sighting tube for a thermal imager, 7 - observation window, 8 - sealed adapter.

2.3. Calculation of the catalytic recombiner parameters
The release of heat on the catalyst granules in the heating unit leads to an increase in air temperature in accordance with the equation [43]:

$$\frac{E_h(n-n_0)}{N_A}vS = \frac{n}{N_A}vSC_{pa}(T-T_0) \quad (1)$$

where N_A is the Avogadro constant with an exact value of 6.02214076×1023 mol^{-1} (unit of reciprocal moles), E_h is the enthalpy of combustion of 1 mole of hydrogen (240 kJ/mol); n-n_0 is the difference in hydrogen concentrations before and after the recombiner; n is the gas concentration at the recombiner outlet; $C_p \cong 7/2$ R is the heat capacity of the mixture at constant pressure, equal to approximately the heat capacity of air, since the percentage of hydrogen is insignificant, v is the air flow velocity, S is the cross-sectional area of the cylinder with the catalyst. If we translate the concentration of hydrogen into volume percentages C, then from (1) the equation for the increment of the flow temperature is as follows:

$$T = T_0 + \frac{E_h}{C_p 100}(C - C_0) = T_0 + T_a \Delta C \quad (2)$$

here T_a=82 K/vol.% is the temperature characterizing the increase in the temperature of the gas mixture during combustion on the catalyst of 1 volume percent of hydrogen. This assumes no heat loss.

2.4. Calculation of the catalytic heating unit parameters

The calculation of heat transfer processes in the channels of the BP through which the heated air flows is presented. It is assumed that the air at the entrance to the channels contains hydrogen and is heated due to its recombination on the platinum catalyst at the initial section of the entrance to the channels.

On the walls of the channel, a heat exchange process occurs, as a result of which the air is cooled, and the BP warms up. The distribution of air temperatures T_g and BP temperature along the channels T_m is described as follows [25,43]:

$$\begin{cases} \frac{\partial T_m}{\partial t} = \frac{\beta_{gm}}{C_m} \int_0^L \left(T_g(x) - T_m\right) dx \\ \frac{\partial T_g}{\partial x} = -\frac{\beta_{gm}}{C_{pa} Q_a} \left(T_g(x) - T_m\right) \end{cases} \quad (3)$$

where β_{gm} is the heat transfer coefficient between the walls of the BP channel and the heated air; L is the length of BP channels; C_{pa} is the heat capacity of air at constant pressure; C_m is the heat capacity of the BP; Q_a is the air flow.

It is assumed that the BP is well thermally insulated from the external environment and has a constant temperature.

Solution of equations (3):

$$\begin{cases} T_g(x) = T_m + T_{gmax} exp\left(-\frac{\beta_{gm}}{C_{pa} Q_a} x\right) \\ \frac{\partial T_m}{\partial t} = T_{gmax} \frac{C_{pa} Q_a}{C_m} \left(1 - exp\left(-\frac{\beta_{gm}}{C_{pa} Q_a} L\right)\right) \end{cases} \quad (4)$$

where T_{gmax} is the maximum temperature of hydrogen - air mixture $T_{gmax} = T_m + \frac{x_H \xi_H \varepsilon_H}{C_{pa}}$; x_H is the hydrogen concentration in the inlet air flow (relative units); ξ_H is the degree of hydrogen combustion (relative units); ε_H is the amount of heat evolved when 1 cm^3 of hydrogen burned ($\varepsilon_H \approx 11$ $J \cdot ncm^{-3}$).

The second part of the solution of equations system (4) is an expression that characterizes the rate of BP heating. As expected, the maximum BP heating rate is formally observed at an infinite flow rate, if the degree of hydrogen combustion is not considered:

$$\frac{\partial T_m}{\partial t} = T_{gmax} \frac{\beta_{gm} L}{C_m} \quad (5)$$

An important value is also the value of the BP heating efficiency η_{rec}, which is equal to the fraction of the heat of hydrogen recombination, which is spent on heating the BP. This value is calculated as follows:

$$\eta_{rec} = 1 - exp\left(-\frac{\beta_{gm}}{C_{pa} Q_a} L\right) \quad (6)$$

When analyzing the effect of air flow on the heating efficiency and rate, it should be noted that in transient flow mode in the channel (from laminar to turbulent), the heat transfer coefficient β_{gm} depends on the air flow through the Nusselt number Nu as follows:

$$\begin{cases}\beta_{gm} = Nu\frac{\lambda}{d} \\ Nu = 0.023Re^{0.8}Pr^{0.4}\end{cases} \tag{7}$$

where *Re* is the Reynolds number; *Pr* is the Prandtl number; α is a constant independent of air flow.

Thus, analyze the described equations it can be finally conclude that with an increase in air flow increases the heating rate of the BP and decreases the heating efficiency BP (6). With an increase in the hydrogen concentration in the air, the heating rate decreases, which requires an increase in air flow (3) and causes an increase in pressure losses. Based on equations (4)-(6) the requirements for high efficiency of BP heating (low air mixture flow rate) lead to an increase in the BP heating time. To reduce the heating time of the BP while maintaining high efficiency the hydrogen concentration should be maximized and the diameter of the recombiner reduced or increased their length L (5). However, an increase in the hydrogen concentration above 4% vol. unsafe due to the risk of hydrogen igniting when mixed with air. A decrease in the internal diameter of the recombiner leads to a nonlinear increase in gas-dynamic resistance and energy consumption for compression of the hydrogen-air mixture. The same, although to a lesser extent, is correct as the length of the recombiner channels increases.

2.5. Assembling of PEMFC stack

In the work, a demonstration setup of a mobile power supply based on PEMFC short stack (3 cells) with a rated power of 50 W was used. For the manufacture of module components and assembly of cells and stack, the techniques presented in detail in [7, 44] were used. The working area of a rectangular electrode (5.0×8.5 cm^2) was about 42.5 cm^2. Main components of MEAs included:

- Membrane - Nafion™ 212
- Anode - carbon cloth ELAT LT 1400, coated with a catalytic layer based on Pt/C[10] (40 wt. % of Pt on Vulcan XC-72 with 10 wt. % of PTFE DISP 30 Fluoropolymer Dispersion), catalyst loading was 0.5 mg/cm^2, ionomer content in the catalytic layer was about 15 wt.% (dispersion D1021 Nafion™ Dispersion - Water based 1100 EW at 10 wt%)
- Cathode - carbon cloth ELAT LT 1400, with a catalytic layer based on Pt/C (40 wt. % of Pt on Vulcan XC-72), catalyst loading was 0.5 mg/cm^2, ionomer content in the catalytic layer was about 15 wt.%.

MEAs are placed between BP with a build-in catalytic heating unit. An open-cathode scheme was chosen as the design of the PEMFC stack. Fig. 3 shows the general view of the module and the assembly steps of the MEA cathode inside the PEMFC stack (Fig. 3 (1,2,3)). BPs tightened with six studs.

The purpose of the BP is to maintain compressive stresses in the PEMFC to ensure contact between the electrodes and the membrane, as well as to pressurize the adhesive layers of the frames. In addition, the BP serves to fasten the hydrogen supply fittings. Finally, BPs act as a catalytic hydrogen recombiner for the cold start procedure of the PEMFC stack.

Figure 3. Image of the PEMFC stack in a heat-insulating covering with a built-in catalytic unit and the process of PEMFC cathode assembling (1,2,3).

2.6. Testing a module with an embedded catalytic heating unit in conditions of subzero temperatures
The PEMFC stack was frozen in a Pozis (Paracels) freezer. The PEMFC stack was placed in a freezer and cooled down to –40 °C. After freezing the module was connected to the laboratory test setup.

The stack was heated by using the heat from the combustion of hydrogen in the air flow in the catalytic unit to the given temperature in the range from 20 to 80 °C (the temperature in the freezer was maintained at –40 °C). Monitoring of the characteristics, namely the temperature and current density of the stack at a cell voltage of 0.5 V, was carried out every 3 minutes until the temperature reached 0 °C, then every 15 minutes.

Fig. 4 shows a diagram of the experimental setup. PEMFC heating is carried out at a constant temperature with the help of a catalytic heating unit embedded in the BP. To do this, hydrogen from the H_2 tank and air from the compressor are fed into the gas mixing container, and then the gas mixture (a mixture of hydrogen with a concentration of 0.2-3.0 vol.% with air at a constant flow rate of 150 mL/min) is fed into the catalytic recombiners, which ensures heating of the BP of the PEMFC stack. An ATN-8060 electronic load (Atacom, Russia) is connected to the electric contacts of the PEMFC. The temperature in the MEA volume is recorded using a thermocouple.

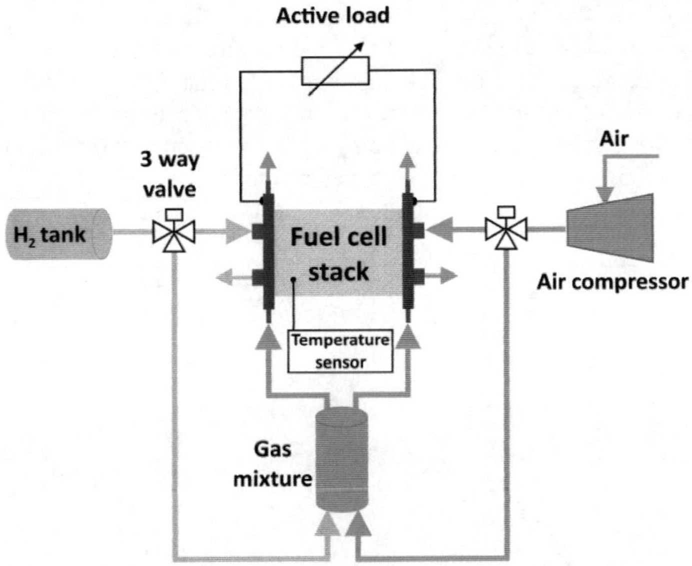

Figure 4. Diagram of a laboratory test system.

The *i-V* curves were recorded using an ATN-8060 electronic load in the potentiodynamic mode in the voltage range from 0.9 to 0.1 V (sweep rate 4 mV/s). The working gases were supplied without humidification at atmospheric pressure and ambient temperature (–40 °C). The flow rates of working gases were: 160 mL/min for hydrogen, and 500 mL/min for air.

The electrochemical impedance spectroscopy (EIS) data was recorded using an electrochemical workstation CorrTest CS350 (CorrTest, China) with an impedance module at a signal amplitude of 10 mV in the frequency range from 50 kHz to 0.1 Hz. The Nyquist plot was recorded at a voltage of 0.5 V. All measurements were carried out with an identical assembly and the same preparation procedure and compares the obtained data with each other.

3. Results and discussion

3.1. Heating module specifications
Fig. 5 and 6 shows the results of optimization the operation of the catalytic hydrogen recombiner. Analysis of thermal images (Fig. 5a) provided to obtain the distribution of temperature increase along

the height of the recombiner depending on the hydrogen concentration in the input flow (Fig. 5b), as well as the average temperature of the recombiner.

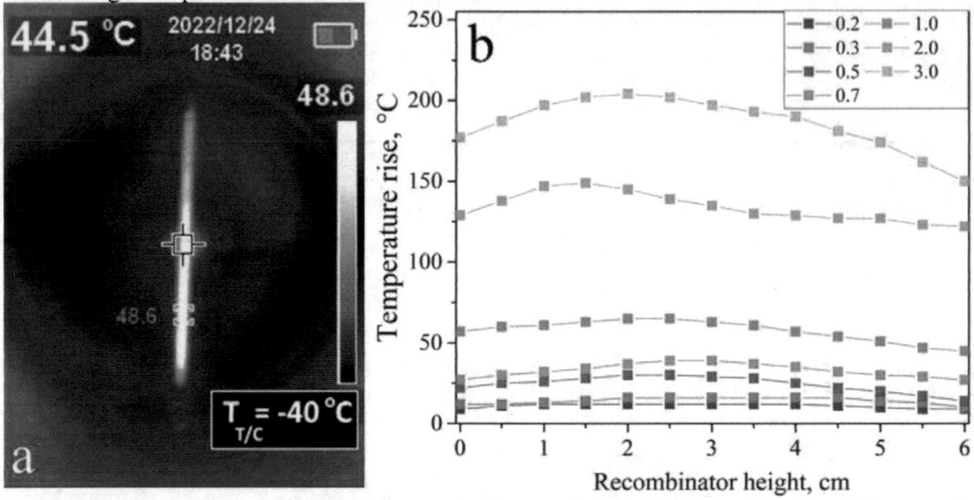

Figure 5. (a) Thermal image of the catalytic recombiner at 5 minutes after the start of the reaction when entering the stationary oxidation process (hydrogen concentration in the air flow is 1.5 vol.%). Lower temperature limit was measured by a thermocouple. (b) Distribution of temperature rise along the height of the recombiner depending on the hydrogen concentration in the input flow.

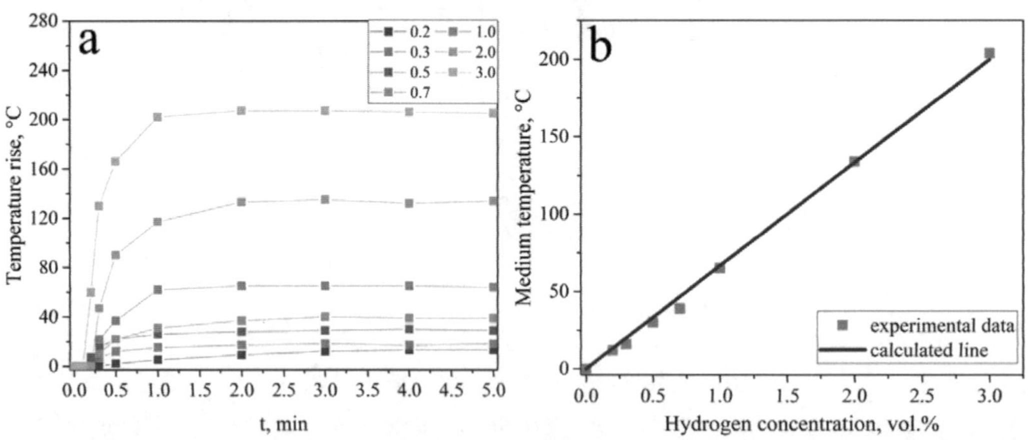

Figure 6. (a) The dependence of the heating time of the recombiner to a given temperature. (b) The dependence of the increase in the temperature of the combustion chamber on the concentration of hydrogen in the input flow, obtained experimentally and calculated according to the presented equations (1) - (2).

The heating time of the BP depends on the hydrogen concentration in the air flow and decreases with increasing concentration from 4 minutes at a concentration of 0.2 vol.% to 1.5 minutes at a concentration of 3.0 vol.%. The temperature characterizing the increase in heat of the gas mixture during combustion of 1 volume percent of hydrogen on a catalyst in a closed system is about 82 °C. In our experiment, the characteristic temperature according to linear interpolation is about 68 °C and increases with an increase in the hydrogen concentration in the flow. The heat utilization efficiency is about 85%, which is a consequence of heat loss through the chamber walls and through radiation. The efficiency of a single recombiner can thus reach 90%. However, the use of a built-in heating unit (consisting of three recombiners) will lead to larger heat losses. The use of the calculation model of the BP heating provided to estimate the efficiency of heating at 60%.

Using the obtained dependences, the conditions for heating the PEMFC assisted by the built-in catalytic heating unit from a temperature of –40 °C to a cell operating temperature of 35 °C were selected.

3.2. Cold start of PEMFC

Hydrogen and air flows were supplied at –40 °C (air from the cold environment) and a relative humidity of no more than 20%, which was affected the efficiency of the PEMFC stack (Fig. 7a). The maximum power received was 0.12 W/cm^2 or 15 W from the PEMFC stack *vs* 50 W rated power. Activation losses (Fig. 7b) are observed both in the anode and pronounced in the cathode spaces, which is caused by the low temperature and humidity of the input gases, as well as the use of an open cathode for air supply.

Fig. 7 shows the *i-V* curves and the Nyquist plot of the PEMFC before freezing.

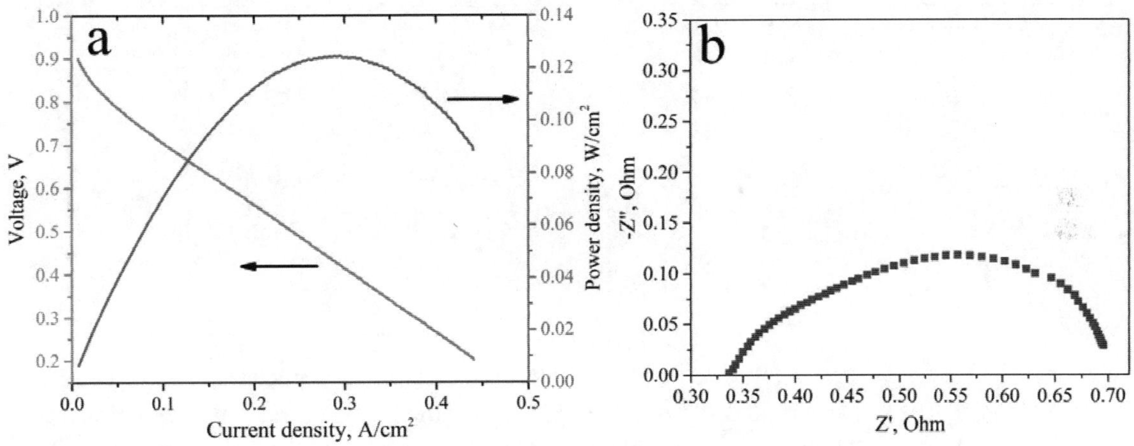

Figure 7. *i-V* curves (a) and the Nyquist plot (b) of the PEMFC before freezing.

Fig. 8 shows the characteristics of the cold start of the PEMFC when the BP is heated to 20, 40 and 60 °C, namely, the dependences of the current density and temperature of the PEMFC on time, and *i-V* curves before freezing and after cold start. The stack is started up in the maximum power mode at a voltage of 0.5 V, which provided the optimal mode for reaching the operating characteristics [25, 45]. The module was heated by a built-in heating unit, a hydrogen-air mixture was used with a hydrogen concentration in the flow of 1.2, 1.8 and 2.2 vol.%, respectively, at a BP temperature of 20, 40 and 60 °C.

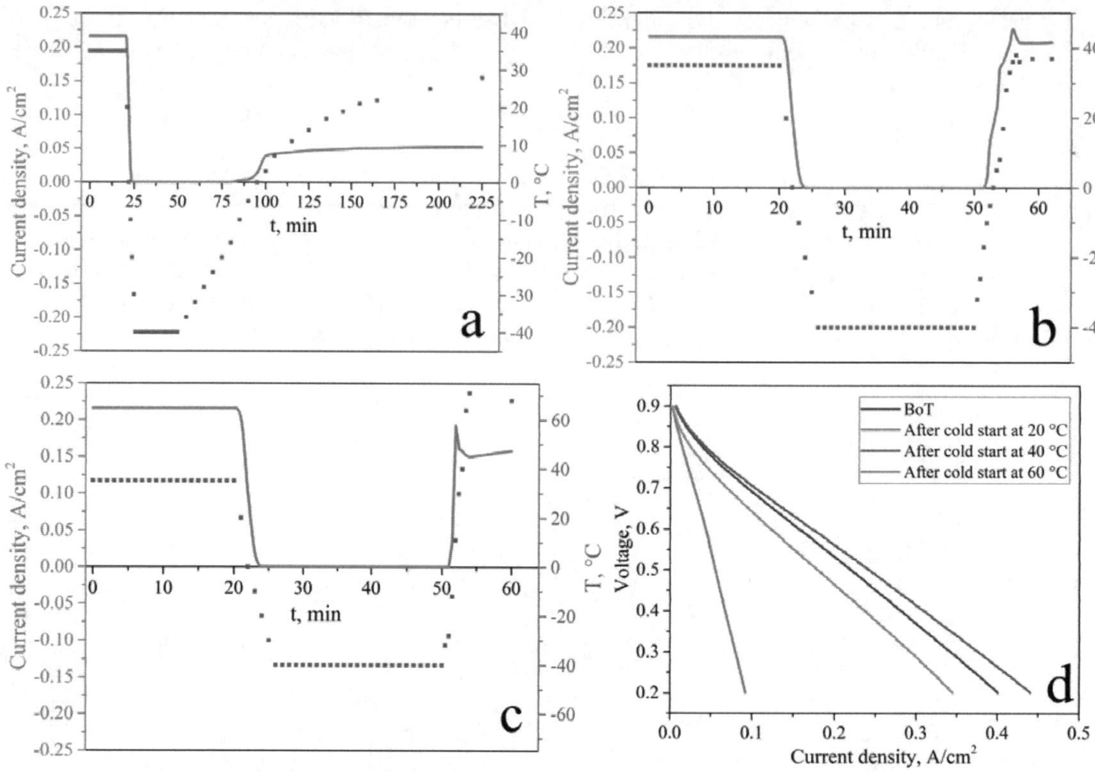

Figure 8. Dependence of the current density and temperature of the PEMFC stack on time during a cold start (a, b, c) and the *i-V* curves after a cold start (d) when the end plates are heated to 20 °C (a), 40 °C (b) and 60 °C (c).

As can be seen from Fig. 8a, heating the BP to 20 °C is not sufficient for a successful cold start, since further self-heating of the PEMFC is not sufficient to reach operating temperatures. Excessive startup time leads to irreversible degradation processes of the MEA components. Based on the results of the EIS, after the start of the PEMFC under conditions of subzero temperatures, the resistance increased significantly, which is associated with the intensive degradation of the MEA, the drying of the membrane, the destruction of the catalytic layer and the disruption of the three-phase interface ionomer-catalyst-support. The PEMFC was defective and did not work, thus cold start failed. This effect is caused by the duration of the heating process due to the low temperature of the heating block. The total warm-up time was more than 2 hours.

Fig. 8b shows the dependence of the current density and temperature of the PEMFC stack on the time after freezing and heating to the operating temperature when the BP are heated to 40 °C. The total warm-up time was 6 min. Uniform temperature distribution over the area of the BP ensures startup safety without local overheating and thermal destruction of the MEA components. As a result, the cold start was carried out successfully, the initial values of the output power of the PEMFC were reached, and the water balance of the membrane and catalytic layers was maintained. In this case, the hydrogen flow was 45 cm^3/s. Hydrogen concentration in the air flow was about 1.8 vol.%.

Fig. 8c shows the current density and temperature of the PEMFC as a function of time during a cold start when the BP are heated to 60°C. The total startup time was 1.5 minutes. A significant drop in current density is noticeable, since a rapid increase in temperature in the PEMFC stack causes significant dehydration of the ionomer, which leads to an increase in resistance of the electrocatalytic layer and of the membrane. The output power of the PEMFC decreased, which indicates the shortcomings of this cold start scenario of the stack. Uncontrolled overheating of the cell is also observed due to self-heating of the stack because of an electrochemical reaction, which will also lead to further degradation of the MEA components.

When comparing the *i-V* curves after a cold start at different heating temperatures, it can be concluded that the scenario is most effective when using close values of the BP temperature and the operating temperature of the PEMFC. In comparison with other options, heating at 40 °C to the operating temperature is carried out rather quickly and at the same time smoothly, provided to control the heating process and prevent dehydration and thermal destruction of the MEAs components. If the operating temperature of the stack is significantly exceeded, the membrane dries up, which leads to a deterioration in the PEMFC performance.

4. Conclusions

The paper proposes a cold start scenario for the PEMFC stack assisted with a heating unit built-in to the bipolar plates. The heating unit is a hydrogen recombiner tube filled with a catalyst. Using of the built-in unit has increased the efficiency of heat transfer up to 60% and reduce the hydrogen concentration in the input flow to 1.8 vol.% in an air flow of 150 mL/min. An increase in air flow leads to an increase in the heating rate and loss of heat transfer efficiency. The hydrogen flow rate is 45 cm^3/s. Assisted cold start processes with build-in catalytic heating unit in proton exchange membrane fuel cell stack has carried out at the start temperature not higher than –40 °C and the parameters for the PEMFC normal operation have determined. Cold start assisted with a heating unit heated to 40 °C was provided the best cold start conditions, including a warm-up time of about 6 minutes and the stability of the PEMFC performance after freezing and subsequent start-up. Fast and smooth heating of the build-in unit and the stack allows controlling the start-up process and ensures the safety of the MEAs components from local overheating and subsequent thermal dehydration and degradation. The proposed method reduces the consumption of hydrogen for heating.

Acknowledgements

The work was carried out within the state assignment of National Research Center "Kurchatov Institute" (Order No. 83), namely investigations of build-in catalytic heater and PEMFC cold start. In part the work was provided within the framework of project FSWF-2023-0014 supported by the Ministry of Science and Higher Education of the Russian Federation, namely analysis of EIS data.

Nomenclature

N_A	Avogadro constant with an exact value of 6.02214076×1023 mol^{-1} (unit of reciprocal moles)
E_h	enthalpy of combustion of 1 mole of hydrogen (240 kJ/mol)
n_0	gas concentration at the recombiner inlet (mol)
n	gas concentration at the recombiner outlet (mol)
C_{pa}	heat capacity of the mixture at constant pressure (7/2 R)
C	gas concentration at the recombiner outlet (vol/%)
C_0	gas concentration at the recombiner inlet (vol.%)
v	air flow velocity (cm^3s^{-1})
S	cross-sectional area of the cylinder with the catalyst (cm^2)
T_0	temperature for the recombiner inlet (K)
T	temperature for the recombiner outlet (K)
T_a	temperature characterizing the increase in the temperature (K/vol.%)
ΔC	difference in hydrogen concentrations (vol.%)
T_g	distribution of air temperatures
T_m	BP temperature along the channels
β_{gm}	heat transfer coefficient
L	length of BP channels (cm)

C_m	heat capacity of the BP
Q_a	air flow (cm^3s^{-1})
T_{gmax}	maximum temperature of hydrogen (K)
x_H	hydrogen concentration in the inlet air flow (relative units)
ξ_H	degree of hydrogen combustion (relative units)
ε_H	amount of heat evolved when 1 cm^3 of hydrogen burned ($J \cdot ncm^{-3}$)
η_{rec}	heating efficiency (relative units)
Re	Reynolds number
Pr	Prandtl number
α	constant independent of air flow

References

[1] Liang J, Fan L, Du Q, Yin Y and Jiao K 2023 Ice Formation during PEM Fuel Cell Cold Start: Acceptable or Not? *Adv. Sci.* **10** 2302151

[2] Patil V, Reshmi P V, Prajna S, Yashaswi, Yashaswini, Haleshappa D, Jayarama A and Pinto R 2023 Degradation mechanisms in PEM fuel cells: A brief review *Mater. Today Proc.* S221478532301670X

[3] Mensharapov R, Ivanova N, Spasov D, Grigoriev S and Fateev V 2022 SAXS Investigation of the Effect of Freeze/Thaw Cycles on the Nanostructure of Nafion® Membranes *Polymers* **14** 4395

[4] Zhan Z, Yuan C, Hu Z, Wang H, Sui P C, Djilali N and Pan M 2018 Experimental study on different preheating methods for the cold-start of PEMFC stacks *Energy* **162** 1029–40

[5] Gießgen T and Jahnke T 2023 Assisted cold start of a PEMFC with a thermochemical preheater: A numerical study *Appl. Energy* **331** 120387

[6] Amamou A A, Kelouwani S, Boulon L and Agbossou K 2016 A Comprehensive Review of Solutions and Strategies for Cold Start of Automotive Proton Exchange Membrane Fuel Cells *IEEE Access* **4** 4989–5002

[7] Ivanova N A, Spasov D D, Grigoriev S A, Kamyshinsky R A, Peters G S, Mensharapov R M, Seregina E A, Millet P and Fateev V N 2021 On the influence of methanol addition on the performances of PEM fuel cells operated at subzero temperatures *Int. J. Hydrog. Energy* **46** 18116–27

[8] Zhu M, Xie X, Wu K, Najmi A-U-H and Jiao K 2019 Experimental investigation of the effect of membrane water content on PEM fuel cell cold start *Energy Procedia* **158** 1724–9

[9] Voloshchenko G N, Zasypkina A A and Spasov D D 2020 Model Study of a Cold Start of a Power Plant Based on a PolymerElectrolyte Membrane Fuel Cells in the Conditionsof Arctic Temperatures *Nanotechnologies Russ.* **15** 326–32

[10] Jeong S, Kim H K, You G J, Im T W and Park Y S 2007 Moisture removal system using adsorbent for a better cold start of FCVs *J. Mech. Sci. Technol.* **21** 1775

[11] Knorr F, Sanchez D G, Schirmer J, Gazdzicki P and Friedrich K A 2019 Methanol as antifreeze agent for cold start of automotive polymer electrolyte membrane fuel cells *Appl. Energy* **238** 1–10

[12] Luo Y and Jiao K 2018 Cold start of proton exchange membrane fuel cell *Prog. Energy Combust. Sci.* **64** 29–61

[13] Li L, Wang S, Wang G, Yue L and Takuto A 2019 Effect of clamping stress on Cold Start Performance of PEMFC *Energy Procedia* **158** 1744–9

[14] Luo Y, Jiao K and Jia B 2014 Elucidating the constant power, current and voltage cold start modes of proton exchange membrane fuel cell *Int. J. Heat Mass Transf.* **77** 489–500

[15] Tabe Y, Saito M, Fukui K and Chikahisa T 2012 Cold start characteristics and freezing mechanism dependence on start-up temperature in a polymer electrolyte membrane fuel cell *J. Power Sources* **208** 366–73
[16] Ishikawa Y, Shiozawa M, Kondo M and Ito K 2014 Theoretical analysis of supercooled states of water generated below the freezing point in a PEFC *Int. J. Heat Mass Transf.* **74** 215–27
[17] Lin R, Zhu Y, Ni M, Jiang Z, Lou D, Han L and Zhong D 2019 Consistency analysis of polymer electrolyte membrane fuel cell stack during cold start *Appl. Energy* **241** 420–32
[18] Lin R, Weng Y, Lin X and Xiong F 2014 Rapid cold start of proton exchange membrane fuel cells by the printed circuit board technology *Int. J. Hydrog. Energy* **39** 18369–78
[19] Li L, Wang S, Yue L and Wang G 2019 Cold-start icing characteristics of proton-exchange membrane fuel cells *Int. J. Hydrog. Energy* **44** 12033–42
[20] Amamou A, Kandidayeni M, Boulon L and Kelouwani S 2018 Real time adaptive efficient cold start strategy for proton exchange membrane fuel cells *Appl. Energy* **216** 21–30
[21] Li L, Wang S, Yue L and Wang G 2019 Cold-start method for proton-exchange membrane fuel cells based on locally heating the cathode *Appl. Energy* **254** 113716
[22] Zhu M, Xie X, Wu K, Najmi A-U-H and Jiao K 2019 Experimental investigation of the effect of membrane water content on PEM fuel cell cold start *Energy Procedia* **158** 1724–9
[23] Yang Z, Du Q, Huo S and Jiao K 2017 Effect of membrane electrode assembly design on the cold start process of proton exchange membrane fuel cells *Int. J. Hydrog. Energy* **42** 25372–87
[24] Huo S, Jiao K and Park J W 2019 On the water transport behavior and phase transition mechanisms in cold start operation of PEM fuel cell *Appl. Energy* **233–234** 776–88
[25] Guo Q, Luo Y and Jiao K 2013 Modeling of assisted cold start processes with anode catalytic hydrogen–oxygen reaction in proton exchange membrane fuel cell *Int. J. Hydrog. Energy* **38** 1004–15
[26] Luo Y, Jia B, Jiao K, Du Q, Yin Y, Wang H and Xuan J 2015 Catalytic hydrogen–oxygen reaction in anode and cathode for cold start of proton exchange membrane fuel cell *Int. J. Hydrog. Energy* **40** 10293–307
[27] Guo H, Sun S, Yu H, Lu L, Xu H and Shao Z 2019 Proton exchange membrane fuel cell subzero start-up with hydrogen catalytic reaction assistance *J. Power Sources* **429** 180–7
[28] Luo Y, Guo Q, Du Q, Yin Y and Jiao K 2013 Analysis of cold start processes in proton exchange membrane fuel cell stacks *J. Power Sources* **224** 99–114
[29] Du Q, Jia B, Luo Y, Chen J, Zhou Y and Jiao K 2014 Maximum power cold start mode of proton exchange membrane fuel cell *Int. J. Hydrog. Energy* **39** 8390–400
[30] Hiramitsu Y, Mitsuzawa N, Okada K and Hori M 2010 Effects of ionomer content and oxygen permeation of the catalyst layer on proton exchange membrane fuel cell cold start-up *J. Power Sources* **195** 1038–45
[31] Zhou Y, Luo Y, Yu S and Jiao K 2014 Modeling of cold start processes and performance optimization for proton exchange membrane fuel cell stacks *J. Power Sources* **247** 738–48
[32] Patterson T and O'Neill J 2008 PEM Fuel Cell Freeze Durability and Cold Start Project *Report* United States 32-34

Journal of Physics: Conference Series **2723** (2024) 012004 doi:10.1088/1742-6596/2723/1/012004

[33] Wei L, Liao Z, Suo Z, Chen X and Jiang F 2019 Numerical study of cold start performance of proton exchange membrane fuel cell with coolant circulation *Int. J. Hydrog. Energy* **44** 22160–72

[34] Fateev V N, Kulova T L, Seregina E A, Ivanova N A, Spasov D D 2019 Electrochemical technologies for low temperatures *Chem. Probl.* **17** 349–57

[35] Qin Z, Wu K, Wu S, Du Q, Yin Y, Wang B, He S, Zu B, Zhang C, Xi F, Chen W, Miao P, Zhang G, Gong J and Jiao K 2023 Investigation of assisted heating cold start strategies from -40 °C for proton exchange membrane fuel cell stack *Int. J. Green Energy* **20** 1559–72

[36] Voloshchenko G N, Zasypkina A A and Spasov D D 2020 Model Study of a Cold Start of a Power Plant Based on a Polymer Electrolyte Membrane Fuel Cells in the Conditions of Arctic Temperatures *Nanotechnologies Russ.* **15** 326–32

[37] Zeng Z, Bao C, Li C and Ouyang M 2023 A modeling study on water and thermal management and cold startup of unmanned aerial vehicle fuel cell system *eTransportation* **15** 100222

[38] Johnson T A and Kanouff M P 2012 Development of a hydrogen catalytic heater for heating metal hydride hydrogen storage systems *Int. J. Hydrog. Energy* **37** 2304–19

[39] Assarabowski R J, Unkert W T, Bach L A, Grasso A P and Olsommer B C 2004 Method and apparatus for preventing water in fuel cell power plants from freezing during storage

[40] Song S M, Koo I G and Lee W M 2002 The influence of oxygen additions to hydrogen in their electrode reactions at Pt/Nafion interface *Electrochimica Acta* **47** 2413–9

[41] Sun S, Yu H, Hou J, Shao Z, Yi B, Ming P and Hou Z 2008 Catalytic hydrogen/oxygen reaction assisted the proton exchange membrane fuel cell (PEMFC) startup at subzero temperature *J. Power Sources* **177** 137–41

[42] Khandelwal M, Lee S and Mench M M 2007 One-dimensional thermal model of cold-start in a polymer electrolyte fuel cell stack *J. Power Sources* **172** 816–30

[43] Baranov I E, Grigoriev S A, Ylitalo D, Fateev V N, Nikolaev I I 2006 Transfer processes in PEM fuel cell: Influence of electrode structure *Int. J. Hydrog. Energy* **31** 203–10

[44] Ivanova N A, Spasov D D, Mensharapov R M, Kukueva E V, Zasypkina A A, Fateev V N and Grigoriev S A 2023 Efficient and stable subzero operation of a PEM fuel cell with a composite anode using hydrogen-methanol composition during freeze/thaw cycles *Int. J. Hydrog. Energy* **48** 11410–20

[45] Luo Y, Jia B, Jiao K, Du Q, Yin Y, Wang H and Xuan J 2015 Catalytic hydrogen–oxygen reaction in anode and cathode for cold start of proton exchange membrane fuel cell *Int. J. Hydrog. Energy* **40** 10293–307

Risk assessment of fire and explosion accidents in oil-hydrogen fueling station based on fault tree analysis

Lianming Du[1†], Guangrong Liao[3†], Tong Zhang[1], Qian Deng[1], Aijun Li[1,2,*] and Ning Mao[4,5*]

[1]School of Energy and Power Engineering, Huazhong University of Science and Technology, Wuhan 430074, China

[2]China-EU Institute for Clean and Renewable Energy, Huazhong University of Science and Technology, Wuhan 430074, China

[3]Gezhouba Group Transportation Investment Co. LTD, Wuhan 430209, China

[4]Transport Planning and Research Institute, Ministry of Transport, Beijing 100028, China

[5]Laboratory of Transport Pollution Control and Monitoring Technology, Beijing 100028, China

†Contributed equally.

* Corresponding author E-mail address: hust_laj@mail.hust.edu.cn (A.Li); 80523432@qq.com (N.Mao)

Abstract. Owing to the coexistence of gasoline, diesel, and hydrogen, a specific and thorough risk assessment approach must be promptly implemented for oil-hydrogen fueling stations. Consequently, this study makes innovations in considering the safety issues of the joint construction of gas station and hydrogen refueling station, and undertaking the modeling and analysis of fire and explosion incidents at oil-hydrogen fueling stations based on fault tree analysis. The findings reveal that the risk level for fire and explosion at oil-hydrogen fueling stations is categorized as grade III. In our fault tree analysis model, there are 3240 minimum cut sets and 194 minimum path sets. The probability of fire and explosion incidents occurring at oil-hydrogen fueling stations is calculated to be 0.000265. This indicates that while the possibility of such accidents is low, their potential severity is exceedingly high. Ultimately, it is recommended that the safety of oil-hydrogen fueling stations be enhanced through the reinforcement of emergency management, customer safety education, and electrostatic discharge devices.

1. Introduction

As well known, hydrogen energy has been regarded as a clean, efficient and high-density energy with the greatest development potential by more and more countries [1]. Moreover, the integration of transportation and hydrogen energy has become a general consensus to realization the low-carbon development of transportation sector. To guarantee the efficient operation of the hydrogen industry chain, hydrogen infrastructure is particularly important.

Oil-hydrogen fueling station is an important hydrogen energy infrastructure that can solve the problems of land area, construction cost and construction time, and it is a reasonable transition scheme in the early stage of hydrogen energy development [2]. In the oil-hydrogen fueling station, all of gasoline, diesel and hydrogen are easy to burn, explore and leak [3]. When fire happened in the oil pool, the temperature of the flame will have an impact on the hydrogen storage tank [4], then the number of victims of the accident may be much higher than that of the gas station and filling station alone [5]. Therefore, A particular and intensive risk assessment method should be put forward in oil-hydrogen fueling station immediately.

Most of the existing studies focused on the risk assessment of hydrogen refueling stations. For example, Jixin Zhang et al. conducted a quantitative analysis using the DEMATEL methodology to determine the likelihood of unsafe control movements occurring in the hydrogen refueling station system, the study obtained the probability of such occurrences taking place[6]. Ashi Chauhan et al. [7] put forward the integrated BN-BWM-HEART method, which can solve the problems of less data and strong subjectivity in assessing personnel's work reliability. Camila Correa-Jullian et al. [8] used traditional QRA modeling tools, like failure mode and impact analysis, event sequence diagram and fault tree analysis, to sort out the data of liquid hydrogen storage system required for quantitative risk assessment. Park and colleagues [9] evaluated the potential individual and social hazards linked to hydrogen refueling stations through an examination of the safety effectiveness of their protective measures. Haugom et al. [10] analyzed the influence of different leakage sizes on the final consequences based on the Bayesian model.

However, we can find that the evaluation of oil-hydrogen fueling stations seems very few. Especially, the risk factors will become more variable after the joint construction of hydrogen refueling station and oil fueling station. Therefore, this paper calculates the probability of fire and explosion accident, accident risk grade, and analyzes the sensitivity of basic event based on fault tree analysis of oil-hydrogen fueling station, then puts forward suggestions for the safety construction and management of the oil-hydrogen fueling station.

2. System Construction

The oil-hydrogen fueling station is the addition of facilities related hydrogen on the basis of the original oil station or the construction of two sets at the same time. The oil-hydrogen fueling station are divided into 8 parts according to the process function area, such as gas source system, oil unloading system, compression system, gas storage system, oil storage system, gas selling system, oil selling system and auxiliary system [11]. The schematic representation of the oil-hydrogen refueling facility can be observed in Figure 1. The hydrogen gas stored in long tube trailer is discharged to the compression system, and stored in the high-pressure storage tank after pressure increase. The gasoline and diesel are transported to the unloading point by the tank truck, and stored in the oil tank through the unloading system. In addition, a gasoline and diesel recovery system with a certain slope pipeline is set up underground in the oil-hydrogen fueling station.

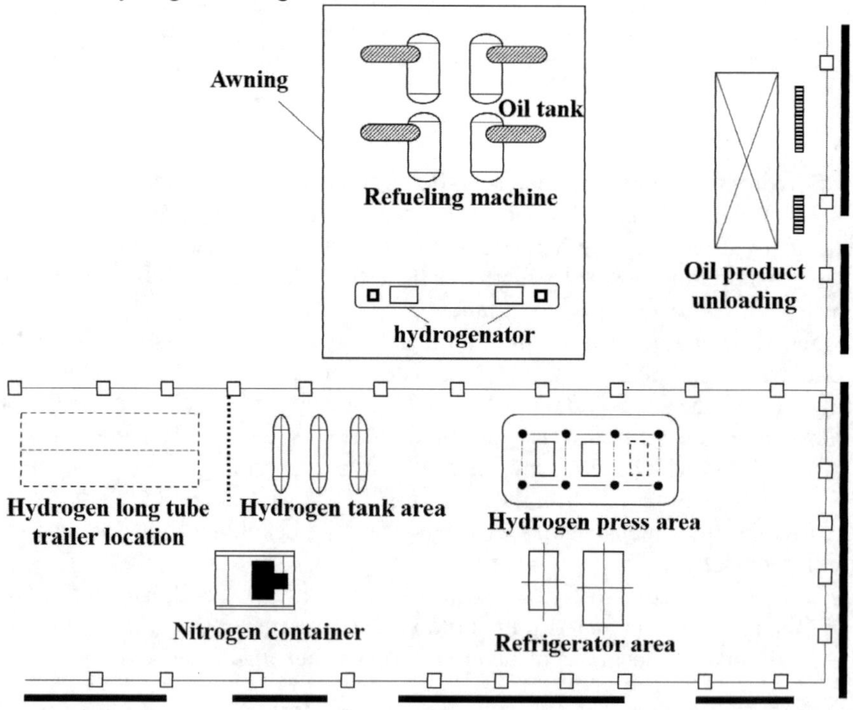

Figure 1. The plane layout diagram of the oil-hydrogen fueling station [12].

The co-construction of hydrogen fueling station and oil station is an efficient way to speed up the popularization of hydrogen energy fuel vehicles. However, on the account of the coexistence of three dangerous chemicals, hydrogen, gasoline and diesel, in the event of an accident, it is crucial to consider not only the explosive combustion mechanism of one fuel substance, but also the interaction between different fuels.

3. Methodology

3.1 Establishment of fault tree

Some scholars have concluded the risk and severity of fire and explosion accidents are the greatest [13]. Therefore, "Fire and explosion accident of oil-hydrogen fueling station" is selected as the top event of this accident tree for analysis. With reference to the cases analysis[14], accident mechanism [15][16] and hydrogen properties [17], the logical connection between the top event and basic events can be established. Then, Fault tree can be figured out (shown in figure 2 and 3). Additionally, the name and probability of each code is shown in table 1.

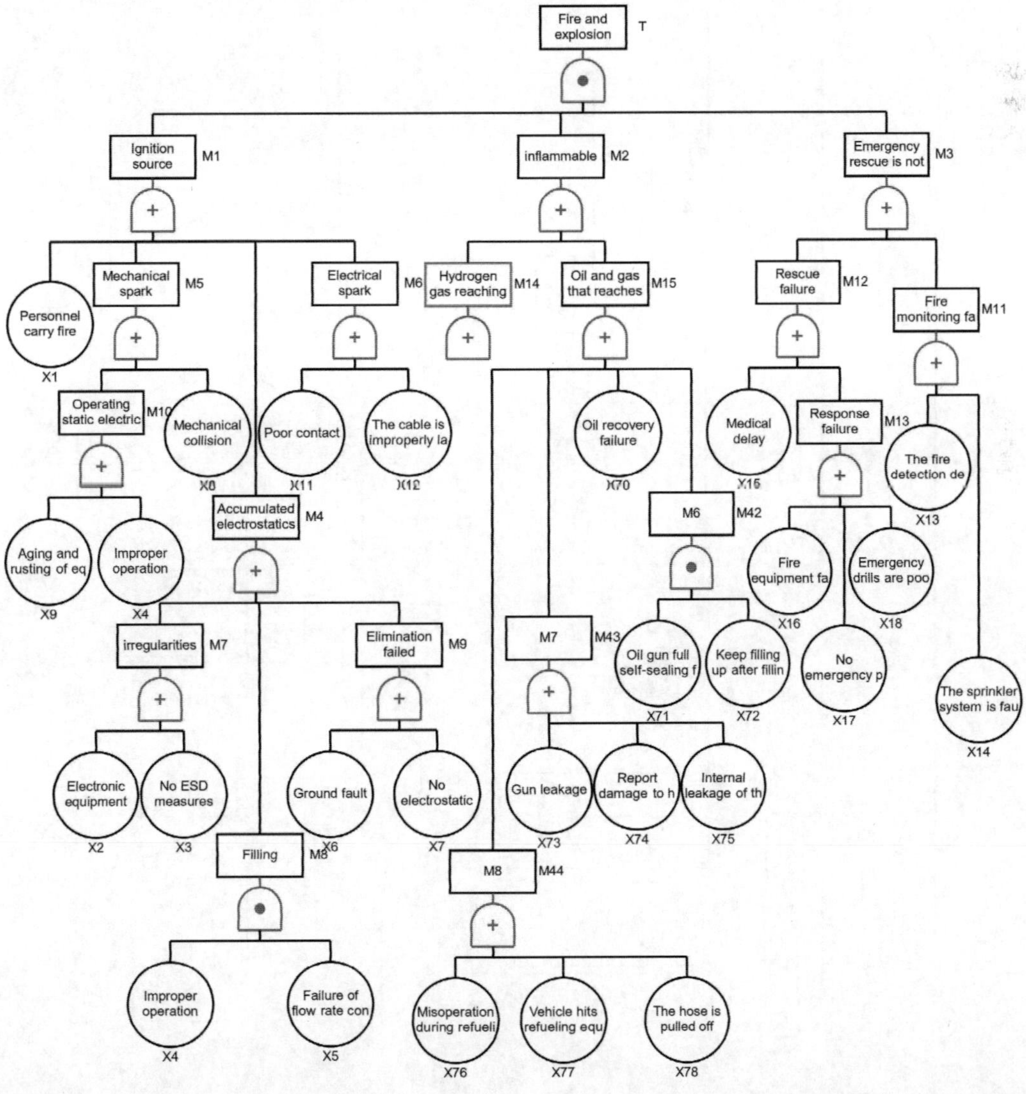

Figure 2. Fire and explosion fault tree in oil-hydrogen fueling station (without detailed M14).

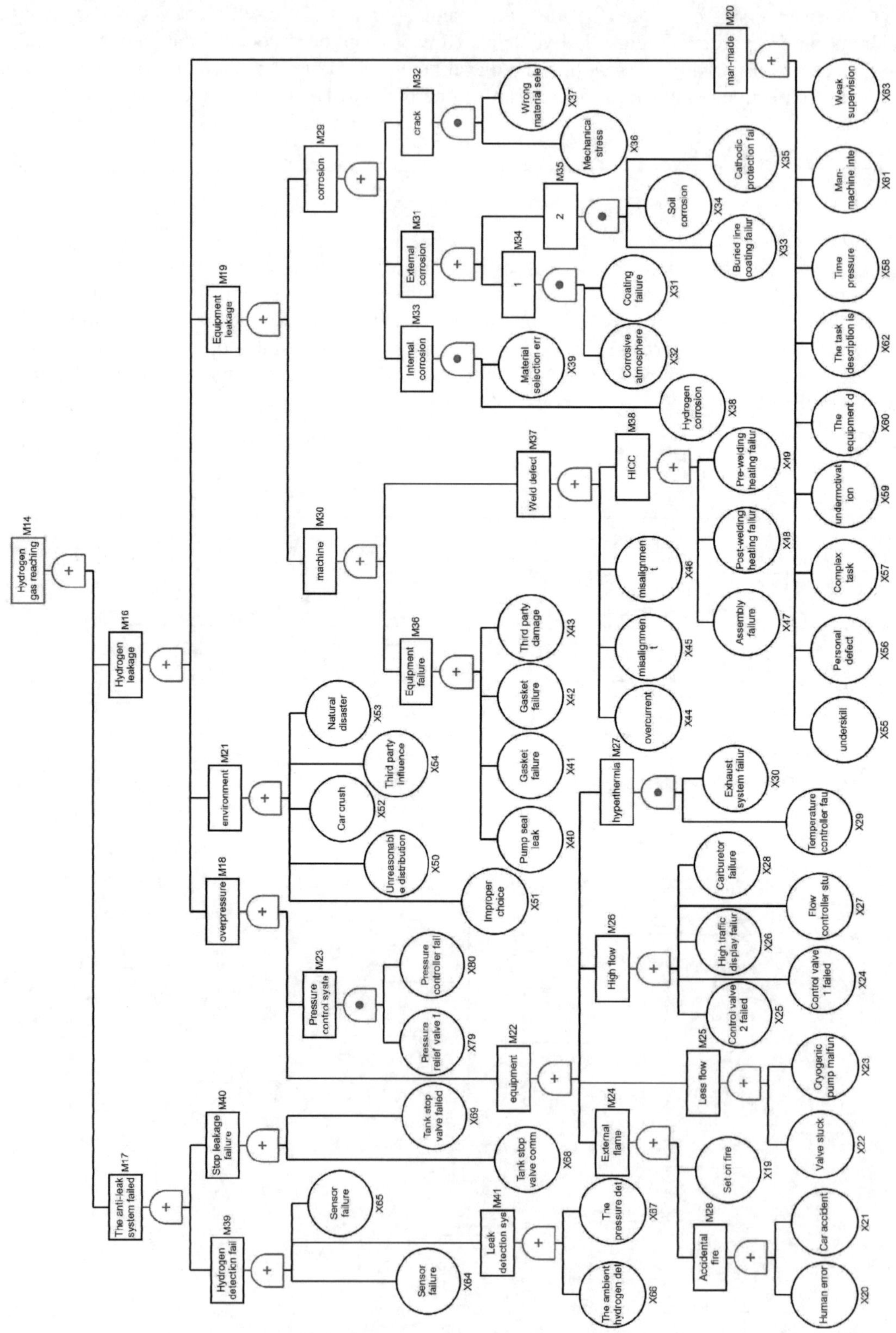

Figure 3. Detailed fault tree of event M14.

Table 1. The possibility of the basic event.

D	P	R	D	P	R	D	P	R
X1	0.00645		X29	0.006195	[13]	X56	0.00755	[19]
X2	0.00496		X30	0.00766		X57	0.00871	
X3	0.0000915		X31	0.000221		X58	0.011	
X4	0.00195		X32	0.001652		X59	0.025	
X5	0.0000511		X33	0.012321		X60	0.0209	
X6	0.000102		X34	0.000175		X61	0.0209	
X7	0.0000406		X35	0.00364		X62	0.0105	
X8	0.00208		X36	0.015963		X63	0.0377	
X9	0.00219	[18]	X37	0.022145		X64	0.00152	[18][20]
X11	0.00119		X38	0.00041		X65	0.000904	
X12	0.00157		X39	0.036297		X66	0.000125	
X13	0.000299		X40	0.007053		X67	0.000068	
X14	0.000138		X41	0.008421		X68	0.00731	
X15	0.00546		X42	0.0000389		X69	0.0253	
X16	0.00177		X43	0.001044		X70	0.000634	[21][22]
X17	0.00526		X44	0.000261		X71	0.000313	[23][24]
X18	0.0122		X45	0.000652		X72	0.00398	
X19	0.067642	[13]	X46	0.002326		X73	0.000199	
X20	0.014746		X47	0.000147		X74	0.000199	
X21	0.064892		X48	0.0897		X75	0.000744	
X22	0.00824		X49	0.001319		X76	0.000744	
X23	0.02717		X50	0.000277	[18]	X77	0.0000228	
X24	0.01074		X51	0.000121		X78	0.000531	
X25	0.011909		X52	0.00087		X79	0.020572	[13]
X26	0.053839		X53	0.000109		X80	0.030028	
X27	0.00437		X54	0.00142				
X28	0.000264		X55	0.0456	[19]			

Notes: D-designation; P-probability; R-reference.

3.2 Minimum cut set

The minimum cut set refers to the minimal combination of basic events that can lead to the occurrence of the top event, specifically the fire and explosion accident at an oil-hydrogen fueling station. If all the basic events within a minimum cut set take place, the top event is guaranteed to occur. The common method for finding the minimum cut set is Boolean algebraic simplification. First, according to the structure of the fault tree, the AND gate is replaced by multiplication, the OR gate is replaced by addition, then the structure function of the top event can be written down successively. The calculation process is shown in the following equation.

$$T=M_1M_2M_3$$
$$=(X_1+M_5+M_4+M_6)(M_{14}+M_{15})(M_{12}+M_{11})$$

••• •••

$=X_1X_{13}X_{19}+X_1X_{13}X_{20}+ \cdots +X_9X_{18}X_{79}X_{80}$

It is calculated that the structure function consists of the addition of 3240 terms, which means this study utilizes Boolean algebra calculations to identify a total of 3240 minimum cut sets associated with fire and explosion accidents at oil-hydrogen fueling stations. The quantity of minimum cut sets directly corresponds to the number of potential pathways leading to the "fire and explosion accidents of oil-hydrogen fueling station."

3.3 Minimum path set

The minimum path set, in contrast to the minimum cut set, refers to the smallest combination of fundamental events that are necessary to prevent the occurrence of the top event, such as a fire and explosion accident at an oil-hydrogen fueling station. If any of the fundamental events in the minimum path set do not take place, the top event will not occur. To identify the minimum path set, it is essential to substitute the AND gate in the fault tree with the OR gate, and the OR gate with the AND gate to create the success tree. Subsequently, the minimum cut set in success tree corresponds to the minimum path set in fault tree.

3.4 Importance

Each basic event within the fault tree will exert varying degrees of influence on the top event.

(1) Structural importance. From the standpoint of fault tree structure, the structural importance denotes the impact on the top event given that the possibility of each basic event is uniform. The structural importance can be calculated utilizing the quadratic approximation method formula, as depicted in equation (1) [25].

$$I_{\Phi}(j) = 1 - \prod_{x_j \in G_r}(1 - \frac{1}{2^{n_j - 1}}) \quad (1)$$

(2) Probability importance. The effect of changing the possibility of a basic event on the top event. Probability importance can be calculated by taking the partial derivative of the probability function of the top event, as represented by equation (2)[25].

$$I_g(i) = \frac{\partial g}{\partial q_i} \quad (2)$$

(3) Critical importance. The critical importance pertains to the rate at which the probability of the top event is influenced by the rate of change in the probability of a basic event, as represented by equation (3)[25].

$$I_c(i) = \frac{q_i}{g} I_g(i) \quad (3)$$

In summary, poor emergency drills, personnel carrying fire, electronic equipment, medical delay and no emergency plan are the five main factors affecting the accident of oil-hydrogen fueling station.

3.5 Probability of the top event

When calculating the possibility of the occurrence of the top event, either the minimum cut set formula or the minimum path set formula can be used. Since there are 3240 minimum cut sets in the fire and explosion fault tree of oil-hydrogen fueling station, the calculation formula of the minimum cut set is complex and time-consuming. Therefore, the calculation formula of the minimum path set is used here, which is show as equation (4)[25].

$$g = 1 - \sum_{r=1}^{l} \prod_{x_i \in P_r}(1 - q_i) + \sum_{1 \le r \le s \le l} \prod_{x_i \in P_r \cup P_s}(1 - q_i) - \sum_{1 \le r \le s \le t \le l} \prod_{x_i \in P_r \cup P_s \cup P_t}(1 - q_i) + \cdots + (-1)^{l} \prod_{\substack{r=1 \\ x_i \in P_r}}^{l} (1 - q_i) \quad (4)$$

4. Result analysis and discussion

By employing tree conversion and Boolean algebra, it has been determined that there are 194 minimum path sets associated with fire and explosion accidents at oil-hydrogen fueling stations. Each of these minimum path sets represents a specific pathway to mitigate the occurrence of the top event. Consequently, there exist 194 distinct pathways for controlling the incidence of "fire and explosion accidents at oil-hydrogen fueling stations". Critical importance, which evaluates the significance of basic events based on sensitivity and probability, is deemed more rational and practical compared to structural and probability importance. Therefore, the subsequent analysis will utilize the results of critical importance as the primary basis, while considering the results of structural and probability importance as supplementary references, to assess the sensibility of each basic event. The findings of this analysis are presented in table 2.

Table 2. The top five basic events in terms of importance.

Event number	Event name	Critical importance ranking	Probability importance ranking	Structure importance ranking
X18	poor emergency drills	1	2	1
X1	personnel carrying fire	2	1	2
X2	electronic equipment	3	1	2
X15	untimely medical treatment	4	2	1
X17	no emergency plan	5	2	1

The aforementioned analysis demonstrates that the probability of fire and explosion incidents at the oil-hydrogen fueling station is 0.000265, with a minimum cut set of 3240 and a minimum path set of 194. The factors with the most significant influence on the occurrence of fire and explosion accidents include poor emergency drills, personnel carrying fire, electronic equipment, untimely medical treatment and no emergency plan.

Based on the available literature, the fault tree analysis of a solitary hydrogen refueling station indicates that the number of minimum cut sets is 144, while the number of minimum path sets is 16 [26]. Upon the concurrent establishment of a gas station and a hydrogen refueling station, it is evident that both the minimum cut set and minimum path set numbers increase. This suggests a greater proliferation of accident occurrence and prevention mechanisms, albeit with a more pronounced rise in accident occurrence pathways. Consequently, it is apparent that the risk level escalates following the joint construction of gas stations and hydrogen refueling stations.

The concept of risk can be expressed as the multiplication of the probability and the severity. Relying solely on fault tree analysis to assess accident probability may not provide a comprehensive evaluation of the risk level associated with an oil-hydrogen fueling station. It is essential to also take into account the potential severity of accidents. Previous studies have indicated that fire and explosion accidents at oil-hydrogen fueling stations pose significant hazards and can result in substantial damage, warranting a classification of very serious severity.

The probability of an accident and the potential impact of its consequences are integrated into the risk matrix table depicted in Figure 4. Analysis reveals that the risk classification of the oil-hydrogen fueling station is situated within region III, indicating a significant level of risk.

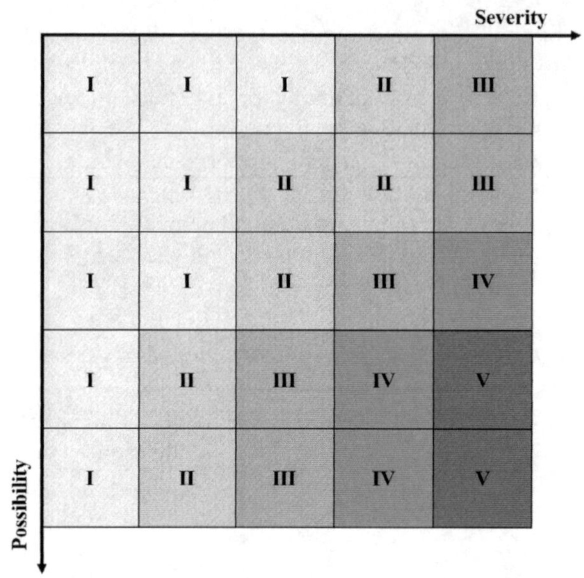

Figure 4. Risk matrix[27].

Several researchers have suggested various approaches to mitigate the risk associated with hydrogen refueling stations. These include the implementation of robust operational protocols, rational scheduling of employees' work hours, and enhanced maintenance of equipment [6]. Additionally, strategies to reduce the risk level of gas stations have been proposed, such as eliminating mobile phone reception, minimizing mechanical collisions, and mitigating container static electricity[22]. Ultimately, to mitigate the risk associated with oil-hydrogen stations, measures can be implemented across following dimensions.

(1) Strengthening emergency management. Strictly implementing the main responsibility of enterprise emergency management, so as to achieve comprehensive coordination, overall management, and special responsibility. Developing a scientific emergency plan, and ensuring that every staff in the station is familiar with the emergency plan. Increasing the investment in emergency management funds, improving emergency equipment and facilities, and regularly carrying out emergency drills.

(2) Strengthening customer safety education. Popularizing the safety knowledge of oil-hydrogen fueling station to customers through slogans, brochures and videos. Carrying out safety training for station staff to ensure correct guidance to customers.

(3) Adding electrostatic discharge device. Electrostatic discharge devices should be set next to high-pressure hydrogen storage tanks, oil and gas storage tanks, hydrogenators and refueling machines. If Electrostatic discharge devices don’t work, the injection cannot proceed, to avoid accidents caused by human static electricity and electronic equipment.

5. Conclusion

This paper study utilized the fault tree analysis method to evaluate the potential for fire and explosion incidents at oil-hydrogen fueling stations. The model generated 3240 minimum cut sets and 194 minimum path sets, ultimately yielding a calculated probability of 0.000265 for fire and explosion accidents. The most influential factors contributing to these incidents were identified as poor emergency drill, personnel carrying fire, electronic equipment, untimely medical treatment and no emergency plan. The analysis findings indicate that the risk level of oil-hydrogen fueling stations is classified as grade III, signifying a low probability of occurrence but with severe potential consequences.

Acknowledgment
This work was supported by "Key R&D Program of China Energy Construction Co. LTD (CEEC2021-KJZX-08-01)".

Reference

[1] Kang J, Wang L, Jin H and Zhang J 2022 Proposed model of potential accident process at hydrogen refueling stations based on multi-level variable weight fuzzy Petri net. *Int. J. Hydrogen Energy* **47**(67) 29162-29171.
[2] Zhang X, Pei W, Mei C X, Deng W, Tan J X and Zhang Q Q 2022 Transform from gasoline stations to electric-hydrogen hybrid refueling stations: An islanding DC microgrid with electric-hydrogen hybrid energy storage system and its control strategy. *Int. J. Elec. Power* **136** 107684.
[3] Vudumu S K 2010 Experimental and computational investigations of hydrogen safety, dispersion and combustion for transportation applications. *Missouri University of Science and Technology*.
[4] Sakamoto J J, Nakayama J, Nakarai T, Kasai N, Shibutani T and Miyake A 2016 Effect of gasoline pool fire on liquid hydrogen storage tank in hybrid hydrogen–gasoline fueling station. *Int. J. Hydrogen Energy* **41**(3) 2096-2104.
[5] Kang J, Wang Z X, Jin H, Dai H Y, Zhang J X and Wang L D 2023 Dynamic risk assessment of hybrid hydrogen-gasoline fueling stations using complex network analysis and time-series data. *Int. J. Hydrogen Energy* **48**(78) 30608-30619.
[6] Zhang J X, Zhang S H, Liang Z W, Lang X S, Shi M H, Qiao J Y, Wei J H, Dai H Y and Kang J 2024 A risk assessment method based on DEMATEL-STPA and its application in safety risk evaluation of hydrogen refueling stations *Int. J. Hydrogen Energy,* **50** 889-902.
[7] Chauhan A, Golestani N, Yazdi M, Njue J C W, Abbassi R and Salehi F 2023 A novel integrated methodology for human reliability assessment in hydrogen fueling stations. *Int. J. Hydrogen Energy,* **48**(34) 12917-12933.
[8] Correa-Jullian C and Groth K M 2022 Data requirements for improving the Quantitative Risk Assessment of liquid hydrogen storage systems *Int. J. Hydrogen Energy,* **47**(6) 4222-4235.
[9] Park B, Kim Y, Lee K, Paik S and Kang C 2021 Risk Assessment Method Combining Independent Protection Layers (IPL) of Layer of Protection Analysis (LOPA) and RISKCURVES Software: Case Study of Hydrogen Refueling Stations in Urban Areas. *Energies,* **14** 4043.
[10] Haugom G P and Friis-Hansen P 2011 Risk modelling of a hydrogen refueling station using Bayesian network *Int. J. Hydrogen Energy,* **36**(3) 2389-2397.
[11] Kwon D, Choi S K and Chulhee Y 2022 Improved safety by cross-analyzing quantitative risk assessment of hydrogen refueling stations *Int. J. Hydrogen Energy,* **47**(19) 10788-10798.
[12] Zhang X Z 2023 Construction design and practice of oil-hydrogen electric joint construction station *Oil depot and gas station* **32**(3) 13-17+4. (in Chinese)
[13] Pirbalouti R G, Dehkordi M K, Mohammadpour J, Zarei E and Yazdi M 2023 An advanced framework for leakage risk assessment of hydrogen refueling stations using interval-valued spherical fuzzy sets (IV-SFS). *Int. J. Hydrogen Energy,* **48**(54) 20827-20842.
[14] Kawatsu K, Suzuki T, Shiota K, Izatoc Y I, Komori M, Sato K, Takai Y, Ninomiya T and Miyake A 2022 Trade-off study between risk and benefit in safety devices for hydrogen refueling stations using a dynamic physical model. *Int. J. Hydrogen Energy,* **47**(57) 24242-24253.
[15] Dagdougui H, Sacile R, Bersani C and Ouammi A 2018 Chapter 7 - Hydrogen Logistics: Safety and Risks Issues. *Hydrogen Infrastructure for Energy Applications,* 127-148.
[16] Astbury G R and Hawksworth S J 2007 Spontaneous ignition of hydrogen leaks: A review of postulated mechanisms. *Int. J. Hydrogen Energy*, **32**(13) 2178-2185.
[17] Molnarne M and Schroeder V 2019 Hazardous properties of hydrogen and hydrogen containing fuel gases. *Process Saf. Environ. Prot.* **130** 1-5.
[18] Xing Y X, Wu J S, Bai Y P, Cai J T and Zhu X P 2022 All-process risk modelling of typical accidents in urban hydrogen refueling stations. *Process Saf. Environ. Prot.* **166** 414-429.

[19] Wang X Y and Gao W 2023 Hydrogen leakage risk assessment for hydrogen refueling stations. *Int. J. Hydrogen Energy,* **48**(91) 35795-35808.
[20] Collong S and Kouta R 2015 Fault tree analysis of proton exchange membrane fuel cell system safety. *Int. J. Hydrogen Energy* **40**(25) 8248-8260.
[21] Liu D, Wang Z Z and Zhang W H 2014 Analysis of fire accident tree at self-service refueling station. *Oil depot and gas station,* **23**(2) 22-25+3-4. (in Chinese)
[22] Ma Q C, Zhang J W and Zhang L B 2014 Accident analysis of fire and explosion in gas station based on improved Bayesian Network. *J. Saf. Sci. Technol.* **10**(6) 176-182. (in Chinese)
[23] Badida P, Balasubramaniam Y, Jayaprakash J 2019 Risk evaluation of oil and natural gas pipelines due to natural hazards using fuzzy fault tree analysis. *J. Nat. Gas Sci. Eng.* **66** 284-292.
[24] Lavasani S M, Ramzali N, Sabzalipour F and Akyuz E 2015 Utilization of Fuzzy Fault Tree Analysis (FFTA) for quantified risk analysis of leakage in abandoned oil and natural-gas wells. *Ocean Eng.* **108** 729-737.
[25] Tong R P, Li G J and Yu C Y 2010 *Common Safety Evaluation Methods and Their Applications* (Bei Jing: China Labor and Social Security Press). (in Chinese)
[26] Liu K, Yang X D and Dai C 2021 Study on the safety risk of hydrogen refueling station based on risk assessment and fault tree analysis. *J. Saf. Sci. Technol.* **17**(S1) 50-55. (in Chinese)
[27] Zhang P, Pan H H and Yang Z Q 2023 Research on risk assessment of urban buried gas pipeline based on risk matrix method and Borda ranking method *J. Saf. Sci. Technol.* **19**(9) 116-122. (in Chinese)

Thermovision as a non – destructive method of technical condition assessment: case study in a smart building

L Dębska[1*], N Krawczyk[1] and A Kapjor[2]

[1] Faculty of Environmental Geomatic and Energy Engineering, Kielce University of Technology, 25-314 Kielce, Poland

[2] University of Zilina, Univerzitna 2, 01026 Zilina, Slovakia

E-mail: ldebska@tu.kielce.pl

Abstract. Currently, more and more emphasis is placed on building energy-efficient buildings from the highest quality materials to keep heat and energy losses as low as possible. Relatively, such buildings are much more expensive than traditional buildings due to the technology that is used in them. This applies to HVAC systems, renewable energy sources and etc. This paper presents the correctness of the thermal insulation of the Energis intelligent building at the Kielce University of Technology with the use of a thermal imaging camera. Measurements and photos were taken outside and inside the building, in the winter, in the late afternoon. This research topic was undertaken because few works and articles offer thermal imaging studies of smart buildings. The analysis of the thermograms showed differences between the surface of the windows and the walls from the inside, and additionally a place in the room in which you can see how cold is trying to penetrate the interior. The study provided data on thermal insulation of the tested building, its assessment and correctness of execution.

Keywords: thermovision, technical condition, building, thermal isolation

1. Introduction

Non – destructive methods of measurements are an important tool for diagnostics and assessment of various elements in many areas of engineering and technology. In particular thermal imaging seems to be of considerable significance due to low cost and ease of use. This technique can be applied for machinery parts' assessment such as drum brakes [1], heat exchangers testing [2 - 5] as well as building applications. In this case thermal bridges can be detected and analysed. Thermal bridges significantly affect the thermal conductivity of external partitions. They also contribute to building defects, such as mold formation, insulation defects and heat loss.

Aghasizadeh et al. [6] focused on assessing the impact of two thermal bridges for reinforced concrete and steel balconies. In their research, 33 different variants of balcony connections were made. In this way, they assessed the linear thermal transmittance of the tested elements and the influence of the temperature coefficient on the external surfaces. More and more often, research on the consumption of thermal energy in buildings is carried out. The use of good thermal insulation is the main aspect of saving thermal energy. On the other hand, other authors [7] performed a study of thermal bridges at the interface between window insulation. Two analyses were performed in the study to determine the thermal conductivity, numerical 2D and 3D. The results showed that the 3D numerical model is more accurate than the 2D model for assessing losses due to thermal bridges. Bergero et al. [8] performed a study on the correction of thermal bridges using a two-dimensional model. The research was carried out in order to check the average values of the heat transfer coefficient. Research on thermal bridges for energy-efficient buildings was also carried out by Cerneckiene et. al. [9].

The topic of air quality parameters, thermal comfort and energy consumption is still an interesting issue, which is linked with proper thermal insulation of buildings. Cakyova et al. [10] in their research focused on the analysis of carbon dioxide concentration in Tseri passive houses in Cyprus and the reduction of CO_2 concentration. The authors showed that the maximum air-tightness of room partitions should be 0.6 air changes per hour. The authors also conducted research in passive houses [11]. The study focused mainly on the assessment of thermal bridges and thermal properties of exterior wall materials. Misiopecki at al. [12] described the most favourable position of the windows in terms of reducing the effect of thermal bridges. Five wall structures and two windows were tested. The results showed that the positioning of the windows influences the amount of energy loss through thermal bridges. The article [13] discusses the assessment of the sustainable built environment and the progress in the sustainable energy industry. On the other hand, another author [14] carried out research on the modernized building using a thermal imaging camera. The results showed, among other things, a lot of irregularities resulting from inadequate thermal insulation of the balcony tides.

The application of infrared systems for the performance of thermal insulation has been successfully done by the authors [15] and other researchers [16] from Kielce University of Technology. During those tests proper settings of the thermal imaging systems have been determined.

The aim of the paper is to conduct a study on the assessment of the technical condition of the sustainable, smart and modern building "Energis" located in the Western part of the campus of Kielce University of Technology using a thermal imaging camera. The paper uses the existing methodology to assess the physical condition of thermal insulation of a smart building in Poland. Such thorough studies conducted in Polish climate conditions, on smart buildings erected according to Polish building regulations, national common practices at the building sites and know-how were not found in literature. Moreover, measurement peculiarities and errors are also discussed in the paper as well as the possibility of using thermovision for thermal resistance and heat flux determination.

2. Selected measurement peculiarities and errors

Thermovision is currently a very common method of temperature determination. It bacame quite cheap and accassible to ordinary users. However, the knowledge of the errors and measurement limitations is quite limited. Conequently, many operators of the thermal systems might not produce correct results of thermal maps that they generate during the measurements.

In order to generate proper temperature readings a lot of parameters need to be input into the camera (or the software). These parameters typically include: emissivity of the observed object, temperature of the air, temperature of the surrounding elements, relative humidity and distance between the camera and the observed object. Proper determination of these parameters enables to obtain correct temperature readings during the tests. However, the most important parameter that requires particular attention is the emissivity value. It ranges from 0 to 1 and should typically be determined with the help of contact temperature measurement systems. Naturally, emissivity of various substances (surfaces) can be found it textbooks, but it is recommended to make individual measurements of the observed element (e.g. concrete wall), because there might be significant differences between the literature data of emissivity and real measurement results.

An example of the importance of proper setting of the emissivity value has been presented in Figure 1. Two cases have been considered: measurements inside of the "Energis" building (with the air and surroundings' temperature of 20°C, relative humidity of 40%, distance between the wall and the camera of 15 m) and outside (with the air and surroundings' temperature of 5°C, relative humidity of 80%, distance between the wall and the camera of 100 m). The temperature readings from the infrared system have been obtained for difference values of emissivity – ranging from 0.51 to 0.99 (the correct value was 0.95). As can been seen the improper values of emissivity lead to significant errors in temperature determination – of a few degrees in the analysed range of emissivity changes. What is worth noticing is that in the case of the measurements taken inside the building, the temperature readings decrease with increasing emissivity (from about 31°C to about 26°C). The opposite is true when the measurements are taken outside (at a long distance from the building, at low temperatures and high relative humidity values). In this case the temperature readings obtained with an infrared camera increase as emissivity rises.

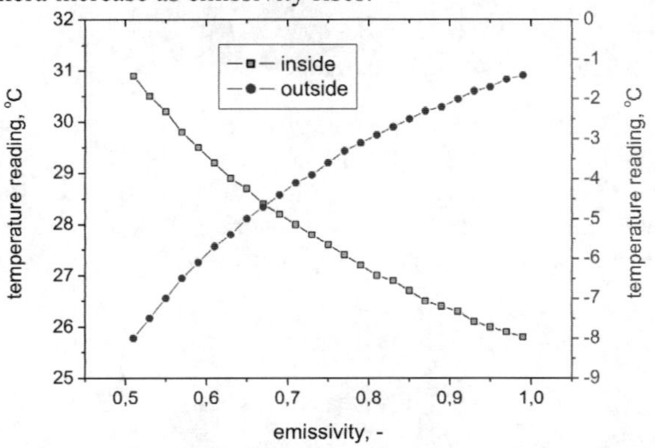

Figure 1. The temperature readings on the inside and outside walls of the Energis building for various emissivity values (winter conditions).

3. Infrared measurements of building elements

Thermovision uses the detection of infrared radiation emitted by every object (naturally, the object's temperature should be above zero Kelvin). The generated thermal map carries a lot of information and data regarding the condition of the surface which undergoes observation. However, care is needed when thermal images are analyzed. This is related not only to the proper determination of input parameters (as presented in section 2), but also to the various phenomena that can affect the readings and lead to misinterpretation of the thermal maps (such as radiation reflection on smooth surfaces, absorption by water vapour and carbon dioxide). Therefore, a thermal analysis of the Energis intelligent building inside and outside during the winter period was undertaken and discussed. The building was completed in 2012 and has a total floor area of 6289 m^2 and cubature of 21211 m^3, while the U-value of the external wall amounts to 0.17 W/m^{2}K. The construction material of the external walls was a 25 cm ceramic hollow brick. There are 2 underground and 5 aboveground storeys. The building is equipped with mechanical ventilation with cooling and heat recovery. All the building services are controlled by the Building Management System (BMS).

Accurate emissivity values were introduced into the measuring device, so as to obtain reliable thermal images (of proper temperature values). The entire Energis building has been presented in Figure 2 – as infrared and digital images.

(a)

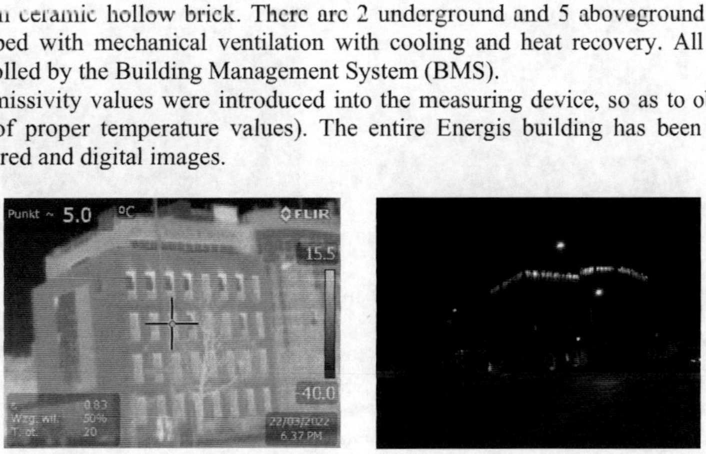

(b)

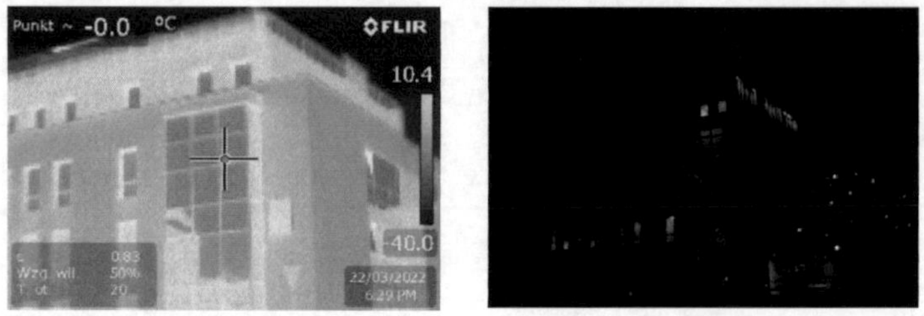

Figure 2. Thermal images of Energis: (a) front facade, (b) rear facade.

Comparing the two thermal images, it can be seen that the building is well insulated. In spite of the fact that it was built about 10 years ago, no defects in its thermal insulation could be detected. There are no thermal bridges, and thus energy losses are not increased. Below in Figure 3 there are photos of the interior of the selected classroom from the inside (room number 3.15 on the third floor).

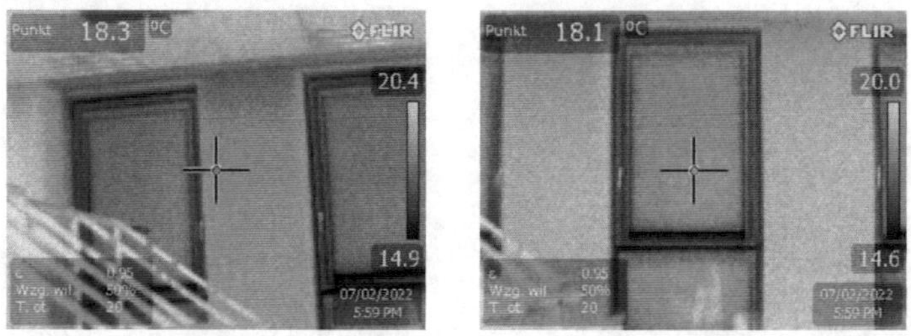

(a) (b)

Figure 3. Classroom 3.15 in the Energis building, from the inside.

Basing on Figure 2 and analyzing the data from Figure 3, it is confirmed that the building has been properly insulated without generating heat and energy losses. No colder spots can occur between the windows, while cooler spots are above the windows closer to the ceiling (Fig. 3 a). This place can be a place where cold air wants to penetrate inside the hall. Figure 4 shows room 3.15 from the outside.

Figure 4. Classroom 3.15 from the outside of the building.

When analyzing the above thermovision image, no construction shortcomings were noticed between the surfaces of the examined windows. Referring to Figure 3, the proper insulation performance of this building is confirmed. However, to further confirm this belief, a photo of the corridor was taken, showing the heat flow in the underfloor heating and a huge window on both sides of the corridor being tested, as shown in Figure 5. Here, it is also worth noting that thermovision cameras could detect any leakage from the underfloor heating, but here there are no defects of the heating system.

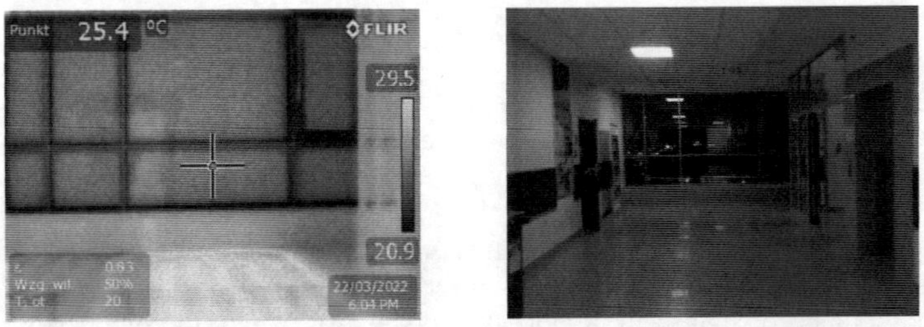

Figure 5. Corridor's thermograph with underfloor heating.

Figure 5 shows the arrangement of heating pipes used for underfloor heating. However, attention should be paid to the huge window with a very large surface. The temperature between it and the wall is significantly different from each other. The reason for such a difference may be the emissivity of each of these elements and the fact that the wall has a much greater thermal resistance than the window. When someone comes close to the window they usually feel cool there, and warmth from the wall.

As seen in Figure 5 thermal imaging can be used for the analysis of the heating systems within the building. In case of any defects (for example in the underfloor heating), the leakage can be easily detected. Figure 6 presents the images (both infrared and digital ones) of the radiator located in a lecture room on the first floor of the "Energis" building. The thermovision technology also enables to determine the hot air movement patterns from the radiators and/or other sources as can be observed in Figure 6a.

(a)

(b)

Figure 6. Thermal imaging of the heating system elements in a lecture room in Energis.

The general and common application of thermovision as a building diagnostic tool is the early detection of thermal bridges and other defect sites. However, it can also be used to help assess thermal resistance of walls (if it is unknown) as well as heat flux dissipated to the surroundings. It can be done through a determination of surface temperature with an infrared camera supported by additional measurements of air temperature with traditional thermometers. Simple heat transfer equations involving thermal conduction (according to the Fourier's law) through walls of known thickness and thermal conductivity value as well as convection on both sides of the walls (according to the Newton's law) can help determine the unknown parameters.

As mentioned earlier in section 2, there are various sources of errors that can influence the readings. Figure 6b illustrates the phenomenon of reflection of thermal radiation emitted from the hot radiators and reflected from a shiny floor and received by the thermovision camera. The floor seems to be hot, but it is just an illusion. This phenomenon should also be taken into account when performing thermovision tests.

Another issue, which is of significant importance, is the influence of thermal performance of walls' thermal insulation on thermal comfort of room users in the buildings. It is mostly related to radiation heat transfer losses from people to the surroundings. The problem of thermal comfort analysis has been considered by the authors (e.g. [17-18]) and might be analysed in the future in conjunction with the thermovision measurements.

4. Summary and conclusions

The following conclusions can be drawn based on data and studies presented in the paper:

- The correct temperature readings with the use of thermovision cameras require proper emissivity values. Otherwise, errors of a few degrees Celcius might occur.
- Possible reflection phenomena should also be considered when analysing the obtained infrared data.
- Thermal imaging studies carried out on a smart building belonging to the Kielce University of Technology proved proper technical condition of the thermal insulation.
- Infrared measurements can be effective in the assessment of thermal resistance of walls (if it is unknown) as well as heat flux dissipated to the surroundings in winter.
- Thermal imaging tests can be an additional source of information, especially for designers to take into account the location of windows and the selection of an appropriate building material.

References

[1] Adamczak S, Orzechowski T and Stańczyk T 2007 *Measur.* **40** (1) 28-35.
[2] Chatys R and Orman Ł J 2017 *Mech. of Com. Mat.* **53** (3) 351-360.
[3] Orman L J 2014 Proc of Int. Conf. "EFM13 - Experimental Fluid Mechanics 2013" (Czech Republic), *EPJ Web of Conferences* **67**, 02087.
[4] Orman L J 2014 Proc of Int. Conf. on Application of Experimental and Numerical Methods in Fluid Mechanics and Energetics (Slovakia), *AIP Conference Proceedings*, 1608, 169-172.
[5] Orman ŁJ and Chatys R 2011, Heat transfer augmentation possibility for vehicle heat exchangers, *Proc. of 15th Int. Conf. "TRANSPORT MEANS" (Kaunas, Lithuania)* 9-12.
[6] Aghasizadeh S, Kari B M and Fayaz R 2022 *Jou. Of Build. En.* **48** 103984
[7] Terentjevas J, Šadauskaitė M, Šadauskienė J, Ramanauskas J, Buska A and Fokaides P 2021 *Cas. Stu. in Con. Mat.* **15** e00768
[8] Bergero S, Cavalletti P and Chiari A 2017 *En. Pro.* **140** p 127-140
[9] Cerneckiene J, Zdankus T, Valancius R and Fokaides P A 2020 *Proc. Of Conf. Earth Environmental Science* **410** 012105
[10] Cakyova K, Figueiredo A, Oliveira R, Rebelo F, Vicente R and Fokaide P 2021 *J. Of Build. En.* **43** 103108
[11] Šadauskienė J, Ramanauskas J, Šeduikytė L, Daukšys M and Vasylius A 2015 *Sus.* **7** 16687-16702
[12] Misiopecki C, Bouquin M, Gustavsen A and Jelle B P 2018 *En. & Build.* **158** 1079-1089
[13] Fokaides P A et. al. 2020 *Sus.* **12** 8417
[14] Stachniewicz R 2018 *Build. Mat.* **12** 37-39
[15] Krawczyk N, Dębska L and Starzomska M 2022 *JoPCS.* **2346** 012010
[16] Stokowiec K, Kotrys-Działak D, Mochocka D and Sokołowski M 2022 *Proc. of The Int. Conf. on Energy and Green Computing (Marocco)* **336**, 00005

[17] Krawczyk N, Dębska L, Piotrowski J Zb, Honus S and Majewski G. 2023 *ROS* **25**, 68-76
[18] Krawczyk N and Dębska L. 2022 *Civil & Environmental Engineering* **18**, 581-588.

Acknowledgements
The experimental study was performed thanks to the funding received from the Polish Ministry of Education and Science.

Supercritical CO_2 recompression Brayton power cycle for hybrid concentrating solar and biomass power plant

C L Liu[1], Z Y Li[1*], N Hao[1], W J Bian[1] and Q Chen[1]

[1]Shanghai Power Equipment Research Institute Co. Ltd, Shanghai200240, China

*Corresponding author. Email: lizhenya@speri.com.cn

Abstract. The supercritical CO_2 Brayton cycle has great potential in various renewable energy systems. The supercritical CO_2 recompression Brayton power cycle for hybrid concentrating solar and biomass power plant is proposed and analyzed in this paper. The supercritical CO_2 is heated by both the molten salt and flue gas from the biomass boiler. The inlet temperature of the turbine in the supercritical CO_2 recompression Brayton power cycle for hybrid power plant rises to 620 °C. The waste heat from the system is recovered by the steam turbine to improve energy utilization efficiency. The efficiency of the supercritical CO_2 recompression Brayton power cycle increases from 0.382 to 0.41. The efficiency of the steam cycle is 0.46, and the efficiency of the combined cycle is 0.427.

1. Introduction

The supercritical carbon dioxide (sCO_2) Brayton cycle drives the turbine with the working fluid CO_2[1]. Owing to the properties of CO_2, like nontoxicity, nonflammability, low critical parameters, the sCO_2 Brayton cycle has a higher efficiency and a simpler configuration than the conventional steam Rankine cycle, and has great potential in various energy systems, like low-grade heat recovery, concentrating solar power (CSP) and fossil fuel power[2-4]. Among these, the solar thermal power (STP) system integrated with molten salt thermal energy storage (MSTES) device and sCO_2 Brayton cycle exhibit flexible output ability and high system efficiency and could play a vital role in future low-carbon energy [5-7]. However, due to the degradation temperature of molten salt (MS) and the intermittent nature of solar irradiation, the STP system integrated with MSTES and sCO_2 Brayton cycle has difficulties in achieving higher solar-to-electric efficiency[8, 9].

Hybridization of renewable energy sources provides a promising solution for the intermittency issues and improvement of the solar thermal energy system[10-13]. Especially, given its relatively steady and reliable output, biomass energy is an attractive option coupling with solar thermal energy. Sahoo et al. [14] study a hybrid biomass-solar power system. In this system, solar thermal energy is used to heat the heat transfer oil and make saturated steam. The inlet temperature of the turbine and energy efficiency rise from 300 °C to 600 °C and from 17.29% to 27.08%, respectively. Anvari et al. [15] found that by adding the solar unit to the biomass power unit, there is a 30% increase in generated power and a 22% decrease in CO_2 emission. Jensen et al. [16] present the design and performance of the hybrid concentrated solar power-biomass plant. The heat provided by a solar collector and biomass boilers is used to generate power using an organic Rankine cycle system, while the waste heat for the local heating. The steam was first generated by the thermal oil and then superheated by the biomass boilers, achieving a higher conversion efficiency. Middelhoff et al. [17] study a hybrid concentrated solar biomass plant involving the biomass boiler and concentrated solar power. The steam can be generated by both the concentrated solar power and biomass boiler. Therefore, the hybrid biomass-solar power generation system performs well in stable power output with high efficiency, especially in raising the steam parameter. However, there are few studies on the hybrid biomass-solar power generation system with sCO_2 Brayton cycle, and the optimization for the hybrid power system is less understood.

The performance of a hybrid power system depends on the configuration optimization of the system and the complementary of different energy sources[18-21]. Wang et al. [22] evaluate the performance of a solar-nuclear power system using the sCO_2 Brayton cycle. Under the condition of insufficient solar energy, the power of the system is only provided by the nuclear power. The mass flow distribution and temperatures of sCO_2 should be changed to match the variation of the system output power. Alenezi et al. [23] analyze the hybrid sCO_2 Brayton power cycle using solar power and natural gas. The heat sources from solar power and oxy-combustion works and heats the hybrid cycle solely. Yang et al. [24] study the load-matching performance of a solar thermal power system with sCO_2 Brayton cycle in a solar-wind hybrid system under load-following mode and stable output mode. The results prove that solar thermal power can provide stable base load for the hybrid system and cooperate well with wind plants. However, there is lack of design for sCO_2 Brayton cycle driven by hybrid biomass-solar power.

Therefore, this paper proposes a sCO_2 recompression Brayton power (RBP) cycle for hybrid concentrating solar and biomass power plant. The thermomechanical analysis of the sCO_2 RBP cycle is conducted. The efficiency of sCO_2 RBP cycle for hybrid power plant is compared with that of the sCO_2 RBP cycle for CSP plants.

2. sCO_2 recompression Brayton power cycle for concentrating solar power plant

The conventional sCO_2 recuperative Brayton power cycle consists of adiabatic compression, isobaric heating, adiabatic expansion, and isobaric cooling. The isobaric heating and isobaric cooling process can be connected by the recuperator, and a simplified reheating cycle is formed, which recover the heat and increases the efficiency of the sCO_2 recuperative Brayton power cycle. Compared to the conventional sCO_2 recuperative Brayton power cycle, the sCO_2 RBP cycle introduces a parallel adiabatic compression process and splits the recuperator into two units, i.e., high-temperature recuperator (HTR) and low-temperature recuperator (LTR). The sCO_2 RBP cycle for CSP plant is shown in Figure 1. The sCO_2 is generated in the main heater (MHE) by the MS, which contains thermal energy from the CSP. The sCO_2 then expands in the turbine. The low-pressure CO_2 output from the turbine will heat the high-pressure sCO_2 in HTR and LTR. After this, the CO_2 will be split into two streams according to the given split ratio. The stream with a higher flow rate enters the precooler (PC), and is in the main compressor (MC), before being heated in the LTR. The other CO_2 is directly compressed in the auxiliary compressor (AC). Finally, the CO_2 streams from the HTR and the LTR mix before the HTR and then are heated in the MHE.

The *T-s* diagram of sCO_2 RBP cycle is shown in Figure 2. sCO_2 is heated up in the MHE by the MS (from state 6 to state 7). Then, sCO_2 expands in the turbine from state 7 to state 8. Then the sCO_2 is cooled in HTR (from state 8 to state 9), and LTR (from state 9 to state 10) sequentially. Exiting from LTR, the sCO_2 splits into two streams. The stream with a higher flow rate is cooled in the precooler (from state 10 to state 1), and is compressed in the MC (from state 1 to state 2), followed by heated in the LTR (from state 2 to state 5). Another stream is compressed in the AC (from state 10 to state 3). The two streams mix before the HTR (from state 5 to state 4 and from state 3 to state 4) and then is reheated in the HTR (from state 4 to state 6), and enter the MHE at state 6 again.

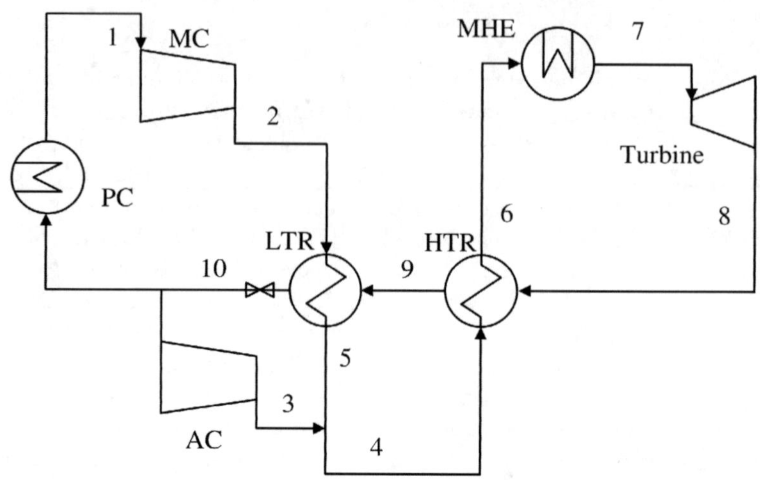

Figure 1. Schematic diagram of sCO_2 RBP cycle for CSP plant.

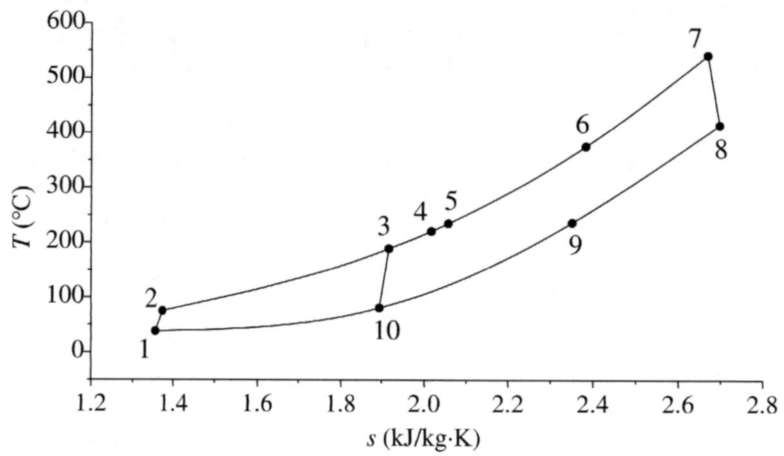

Figure 2. *T-s* diagram of sCO_2 RBP cycle for CSP plant.

To analyze the performance of the sCO_2 RBP cycle, the efficiency of the cycle is:

$$\eta_t = \frac{W_{tur} - W_{com}}{Q_H} \tag{1}$$

$$W_{tur} = m(h_8 - h_7) \tag{2}$$

$$W_{com} = W_{com1} + W_{com2} \tag{3}$$

$$W_{com1} = m_1(h_3 - h_{10}) \tag{4}$$

$$W_{com2} = (m - m_1)(h_2 - h_1) \tag{5}$$

$$Q_H = m(h_7 - h_6) \tag{6}$$

$$r = \frac{m_1}{m} \tag{7}$$

$$\eta_t = \frac{(h_8 - h_7) - r(h_3 - h_{10}) - (1 - r)(h_2 - h_1)}{h_7 - h_6} \tag{8}$$

where, η_t is the efficiency of the sCO_2 RBP cycle, W_{tur} is the energy output of the turbine, W_{com} is the energy consumption of the compressor, Q_H is the heat exchange capacity of MHE, W_{com1} is the energy consumption of the AC, m_1 is mass flow rate of stream with lower flowrate, m is mass flow rate of stream with higher flowrate, W_{com2} is energy consumption of the MC, h_i is the specific enthalpy at each state point in Figure 2, r is the split ratio.

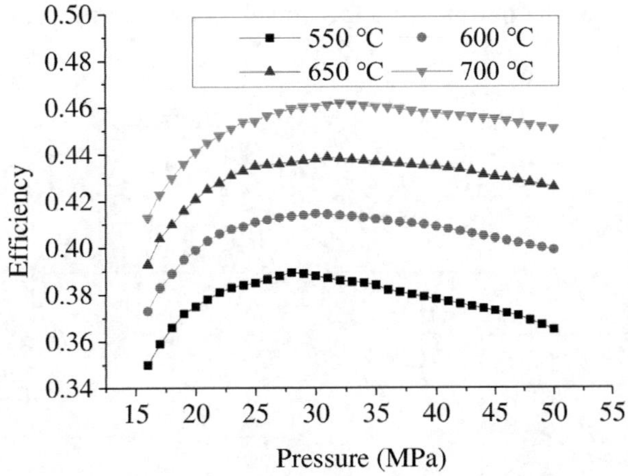

Figure 3. Effects of the inlet pressure and temperature of the 20MW turbine on the efficiency of the sCO_2 power cycle.

Table 1. Parameters and constraints of the sCO_2 RBP cycle for CSP plant.

Parameters		Value
	MC	0.78
Efficiency	AC	0.75
	Turbine	0.90
	MH	0.9 MPa
Pressure loss	HTR	0.2 MPa
	LTR	0.1 MPa
	Precooler	0.05 MPa
Terminal difference	LTR	10 ℃
	HTR	50 ℃
Split ratio		27.9

The parameters and constraints of the sCO_2 RBP cycle are shown in Table 1. To reduce the energy consumption of the compressor and avoid phase change of CO_2, the inlet pressure and temperature a of the MC are 8.47 MPa and 34.5 °C, respectively. The inlet pressure and temperature of the turbine are from 18 to 50 MPa and 550 to 700°C, respectively. An in-house code is used to evaluate the thermodynamic characteristics of the sCO_2 power cycle. The effects of the inlet pressure and temperature of the turbine on the performance of the sCO_2 power cycle are shown in Figure 3. For the same inlet pressure of the turbine, the efficiency of the sCO_2 power cycle increases with the increasing inlet temperature of the turbine. For the same inlet temperature of the turbine, the efficiency of the sCO_2 power cycle rises firstly and then drops with the increasing turbine inlet pressure. As the turbine inlet pressure increases, the compressor operates off optimum condition, and the rise in energy consumption of the compressor is larger than the increase in the energy output of the turbine. Therefore, with the turbine inlet pressure of above 26 MPa, the enhancement of cycle efficiency by increasing inlet pressure of turbine is weak, and, it is better to increase the turbine inlet temperature from 550 °C to 620 °C.

3. sCO_2 recompression Brayton power cycle for hybrid concentrating solar and biomass power plant

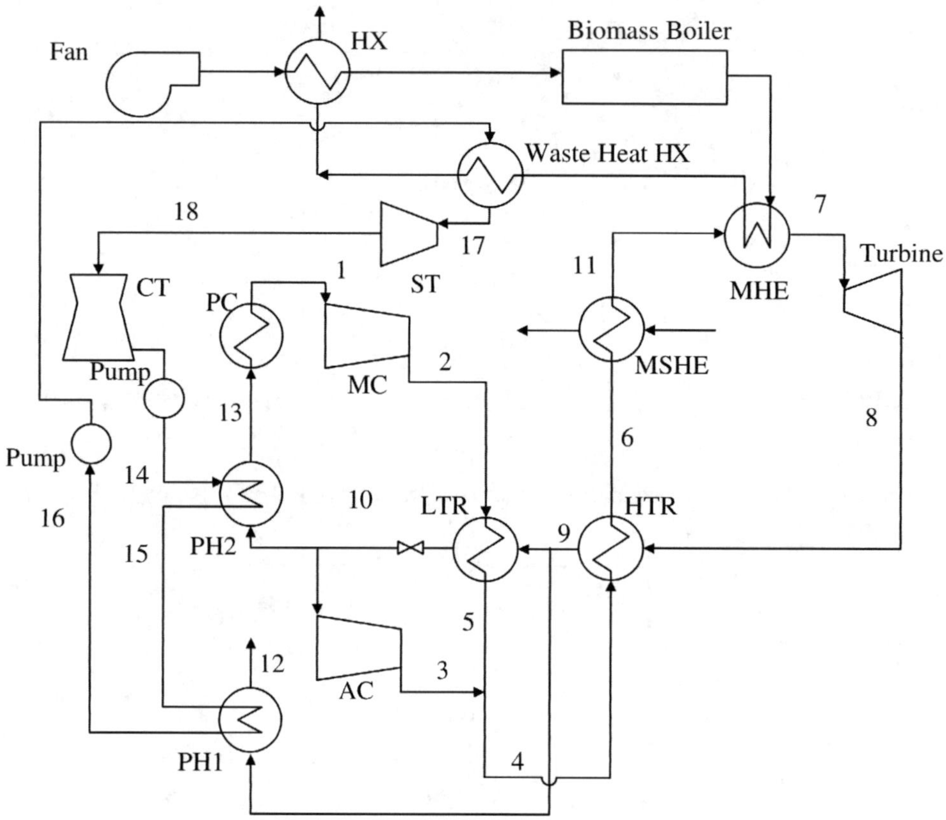

Figure 4. Schematic diagram of sCO_2 RBP cycle for hybrid power plant.

To address the intermittency issue and limitation of degradation temperature of MS, the biomass boiler (BB) is integrated after the MHE. The inter temperature of the turbine can be increased from 550 °C to 620 °C. The waste heat from the BB is utilized by the small steam turbine (ST). Therefore, the efficiency and the economic characteristics of the sCO_2 RBP cycle can be improved. Figure 4 and 5 illustrates the schematic diagram and *T-s* diagram of the sCO_2 RBP cycle for hybrid power plant. Different from the sCO_2 RBP cycle for CSP plant, sCO_2 is first heated up in the MS heat exchanger (HX) (from state 6 to state 11), and then heated in MHE by the flue gas (from state 11 to state 7). Then, sCO_2 expands in the turbine (from state 7 to state 8). Then the sCO_2 is cooled in HTR (from state 8 to state 9). Exiting from HTR, the sCO_2 splits into two streams. The stream with a lower flow rate (stream 1) heats the steam in preheat 1(PH1) (from state 9 to state 12). The stream with a higher flow rate (stream 2) is cooled in LTR (from state 9 to state 10). Exiting from LTR, the sCO_2stream 2 also splits into two streams (stream 3 and 4). The stream with a higher flow rate (stream 3, i.e. main stream) mixes with stream 1 and then is cooled in the preheater 2(PH2) (from state 10 to state 13) and the PC (from state 13 to state 1), and then is compressed in the MC (from state 1 to state 2), followed by heated in the LTR (from state 2 to state 5). All the streams mix before the HTR (from state 5 to state 4 and from state 3 to state 4) and then is reheated in the HTR (from state 4 to state 6), and finally enter the MS heat exchanger at state 6 again. After the MHE, the flue gas heats the steam in the steam generator and the air in the air preheater. The pressure of water from the cooling tower is raised by the pump firstly. Then the water is heated by the

sCO_2in the PH1 and the PH2 (state 14-15-12-16). After going through the water pump, the water is heated by the flue gas in the steam generator, and the steam is generated. The heat in the steam transits to mechanical energy in the steam turbine (from state 17 to state 18). The steam condenses in the cooling tower (CT) finally.

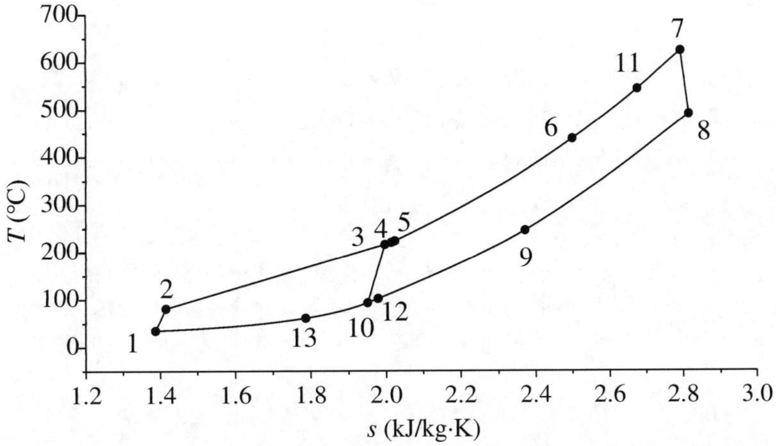

Figure 5. *T-s* diagram of sCO_2 RBP cycle for hybrid power plant.

To analyze the performance of the sCO_2 RBP cycle for hybrid power plant, the efficiency of the cycle is:

$$\eta_t = \frac{W_{tur} - W_{com}}{Q_H} \tag{9}$$

$$W_{com1} = (m - m_1)(h_3 - h_{10}) \tag{10}$$

$$W_{com2} = (m - m_1 - m_2)(h_2 - h_1) \tag{11}$$

$$r_1 = m_1/m \tag{12}$$

$$r_2 = m_2/m \tag{13}$$

$$\eta_t = \frac{(h_8 - h_7) - (1 - r_1)(h_3 - h_{10}) - (1 - r_1 - r_2)(h_2 - h_1)}{h_7 - h_6} \tag{14}$$

where m_1, m_2, and m are mass flow rates of preheater 1, AC and main stream, respectively, W_{com2} is energy consumption of MC, h_i is the specific enthalpy of CO_2, r_i is the split ratio.

The steam cycle efficiency of utilizing the waste heat from the BB can be calculated as

$$\eta_{\mathrm{w}} = \frac{W_{tur.w}}{Q_H} \tag{15}$$

$$W_{tur} = m_{\mathrm{w}}(h_{18} - h_{17}) \tag{16}$$

$$Q_H = m_{\mathrm{w}}(h_{17} - h_{16}) \tag{17}$$

$$\eta_{\mathrm{w}} = \frac{h_{18} - h_{17}}{h_{17} - h_{16}} \tag{18}$$

where η_w is the steam cycle efficiency, $W_{tur,\,w}$ is the energy output of the ST, Q_H is the heat exchange capacity of the steam generator, m_w is mass flow rate of main steam stream, and h_i is the specific enthalpy.

The thermal efficiency of sCO_2 RBP cycle for hybrid power plant can be calculated as follows

$$\eta_{\mathrm{i}} = \frac{Q_{\mathrm{H}} \cdot \eta_{\mathrm{t}} + Q_{\mathrm{w}} \cdot \eta_{\mathrm{w}}}{Q_{\mathrm{H}} + Q_{\mathrm{w}}} \tag{19}$$

where, Q_{H} is the heat input from the MT and BB, η_w is the efficiency of the recompression cycle. Q_w is the input waste heat from the BB. The parameters and constraints of the sCO_2 RBP cycle for hybrid power plant is shown in Table 2.

Table 2. Parameters and constraints of the sCO_2 RBP cycle for hybrid power plants.

Parameters	Value
Pump efficiency	0.78
ST efficiency	0.88
AP Pressure loss	40 KPa
Terminal difference of the preheater 1	10 ℃
Terminal difference of the preheater 1	5 ℃
Split ratio of preheater 2	0.07

According to Figure 5, the efficiency of sCO_2 RBP cycle for hybrid power plant is 41%. The *T-s* diagram of the steam cycle for hybrid power plant is shown in Figure 6. The efficiency of the steam cycle for hybrid power plant is 46%. The efficiency of the combined cycle can reach 42.1%.

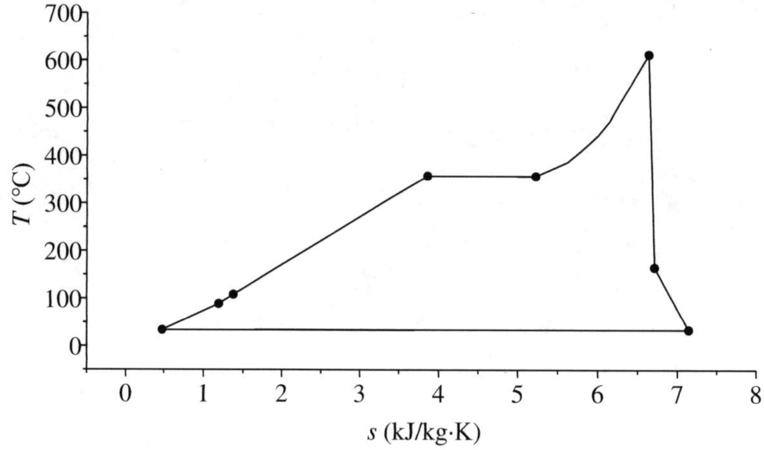

Figure 6. *T-s* diagram of the steam cycle for hybrid power plant.

4. Conclusion

The supercritical CO_2 recompression Brayton power cycle for hybrid concentrating solar and biomass power plant is proposed and analyzed in this paper. The results can be concluded as follows:

(1) Compared with the solar supercritical CO_2 recompression Brayton power cycle, the turbine inlet temperature in the supercritical CO_2 recompression Brayton power cycle for hybrid concentrating solar and biomass power plant rises from 550 °C to 620 °C, and the efficiency of the power cycle increases from 0.382 to 0.41.

(2) The waste heat from the biomass boiler can be recovered by the steam turbine. The waste heat from the supercritical CO_2 can be recovered in the preheater 1 and the preheater 2. The efficiencies of the steam cycle and the combined cycle are 0.46 and 0.427, respectively, which is better than the original ssupercritical CO_2 recompression Brayton power cycle.

References

[1] Ling JLJ, Oh SS, Park HJ and Lee SH 2023 *Renew Sust Energ Rev* **182** 113380.
[2] Fan C, Wu Z, Wang J, Chen Y and Zhang C 2023 *Renew Energ* **210** 810-21.

[3] Hu Z, Fan C, Gao W and Chen Y 2023 *Energy Convers Manage* **293** 117494.
[4] Mubashir W, Adnan M, Zaman M, Imran M, Naqvi SR and Mehmood A 2023 *Therm Sci Eng Prog* **44** 102073.
[5] Zhang C, Li J and Chen Y 2020 *Appl Energ* **259** 114102.
[6] Sun X, Li X, Zeng J, Song Q, Yang Z and Duan Y 2023 *Energy* **283** 128737.
[7] Wu D, Wei M, He J, He S and Tian R 2023 *Appl Therm Eng* **230** 120790.
[8] Lee W, Yoo S, Park DK and Lee K-Y 2023 *Prog Nucl Energy* **163** 104835.
[9] Zhao Y, Chang Z, Zhao Y, Yang Q, Liu G and Li L 2023 *Energy* **282** 128807.
[10] Hu Z, Wang J, Huo E and Zhang C 2023 *Desalination* **559** 116649.
[11] Li Z, Zhu X, Huang X, Tian Y and Huang B 2023 *Sustain Prod Consum* **41** 228-41.
[12] Zhang X, Pan J, Zhao B, Kong X, Ma X, Yi Z, Wang M and Li H 2023 *Appl Therm Eng* **231** 121013.
[13] Fang Y, Li X, Ascher S, Li Y, Dai L, Ruan R and You S 2023 *Energy* **284** 128709.
[14] Sahoo U, Kumar R, Pant PC and Chaudhary R 2016 *SoEn* **139** 47-57.
[15] Anvari S, Khalilarya S and Zare V 2018 *Energy* **165** 776-89.
[16] Jensen AR, Sifnaios I, Perers B, Holst Rothmann J, Mørch SD, Jensen PV, Dragsted J and Furbo S 2022 *Energy Convers Manage* **271** 116207.
[17] Middelhoff E, Andrade Furtado L, Peterseim JH, Madden B, Ximenes F and Florin N 2021 *Energy Convers Manage* **240** 114244.
[18] Zhu J and Chen L 2023 *Appl Energ* **350** 121604.
[19] Li R, Xu D, Tian H and Zhu Y 2023 *Energy* **282** 128706.
[20] Wang M, Shi Z, Luo W, Sui Y and Wu D 2023 *Energy Reports* **10** 1575-88.
[21] Li J, Luo G, Li T, Gao L, Liang X, Hu J, Cao Y, Qi L, Liu X and Huo M 2023 *Energy Strategy Reviews* **49** 101181.
[22] Wang G, Wang C, Chen Z and Hu P 2020 *Energy* **197** 117282.
[23] Alenezi A, Vesely L and Kapat J 2022 *Energy* **247** 123436.
[24] Yang J, Yang Z and Duan Y 2021 *Energy* **223** 120016.

Study of water-side free cooling technology for power lithium battery plant

Jiajing Lin[1,*]

[1]School of Civil Engineering, Chongqing University, Chongqing 400030, PR China

[*]Corresponding author email: 202116021149t@stu.cqu.edu.cn

Abstract. Power lithium battery production process needs to consume a lot of energy. In this paper, an energy consumption analysis of a representative company was conducted, the energy consumption of the facilities that provide support for the process is higher than the energy consumption of the process equipment itself, which greatly affects the cost of battery manufacturing. Air-conditioning system as the most important energy-consuming system of the facilities, accounting for 25% to 45% of the total electricity consumption. The energy consumption of the cooling source system dominates, even in winter its energy consumption can reach 14% of the total energy consumption. On this basis, this paper analyzes the chilled-water-use characteristics of the air-conditioning terminal in winter through theoretical analysis, the chilled water temperature is therefore optimized according to the real-time outdoor humidity. The chilled water temperature is linearly increased after the humidity ratio fall below 7.5g/kg(a), reaching a maximum value of 12.1°C at 6g/kg(a). In addition, the cooling tower free cooling technology based on the optimized chilled water temperature is innovatively proposed to fully explore the energy-saving potential of the technology to achieve the energy saving of the cooling source system. TRNSYS simulation results show that after optimizing the chilled water temperature, the time period for cooling tower individual cooling under winter operating condition is extended by 60%, and cooling source system energy saving rate can reach 49.0% and the cost saving rate 48.0%, both of which are higher than 15% compared to the un-optimized chilled water temperature.

1. Introduction

Lithium battery is characterized by high energy and power density and light weight[1, 2], therefore, it is the first choice for electric vehicle power supply[3].The shipments of power lithium batteries have been growing at a high rate in the past decade, and are expected to reach 1690GWh by 2025[4].

Lithium battery manufacturing cost can be divided into material cost and process cost[5], of which process costs account for 24%. Each process production link is accompanied by huge energy consumption, which can be divided into material upstream energy consumption, production process energy consumption and shell manufacturing energy consumption, accounting for 38%, 38% and 24%, respectively[6]. Among them, the process production energy consumption corresponds to the energy consumed in the battery production plant, which can be further divided into the process energy consumption consumed by each process machine and the energy consumption of plant facilities[7]. Because the battery material is easy to absorb moisture, the workshop needs to strictly control the environmental humidity, which leads to high energy consumption of air-conditioning system[8]. Existing energy-saving studies have focused on rotary dehumidifier, which is the key equipment for creating a low-humidity environment in lithium production workshop[9, 10], while ignore the energy-saving improvement for the cooling source system. This paper will propose an innovative application of a traditional energy-saving measure for the cooling source system to achieve energy saving.

This paper firstly introduces the main process of lithium battery production and the corresponding humidity requirement; then analyzes the energy consumption of the project and the chilled-water-use characteristics of terminals in winter. Due to the lower humidity in winter, the chilled water with too low temperature will exacerbate the cooling and heating offset at the rotor; through analyzing the dehumidification characteristics of the rotor and derives the optimized chilled water temperature according to the real-time fresh air humidity. Finally, free cooling technology based on optimized chilled water temperature is proposed and validates the energy-saving performance of the technology based on the modeling of TRNSYS.

2. Project overview

2.1. Refrigeration system

The project is located in the hot-summer and cold-winter area, China. The main workshop includes pole manufacturing workshop, former process workshop, pole winding workshop, post process workshop, capacity testing workshop, power station and finished product warehouse, with a total of 14 sets of pole production lines, 6 cell production lines, and a planned production capacity of 20GWh per year. There are two sets of chilled water systems, which are installed in the power station and the finished product warehouse. There are 2 low-temperature units and 7 medium-temperature units in power station; and there are 2 medium-temperature units and 2 high-temperature units in finished product warehouse. **Figure 1** provides an overview of the chilled water system.

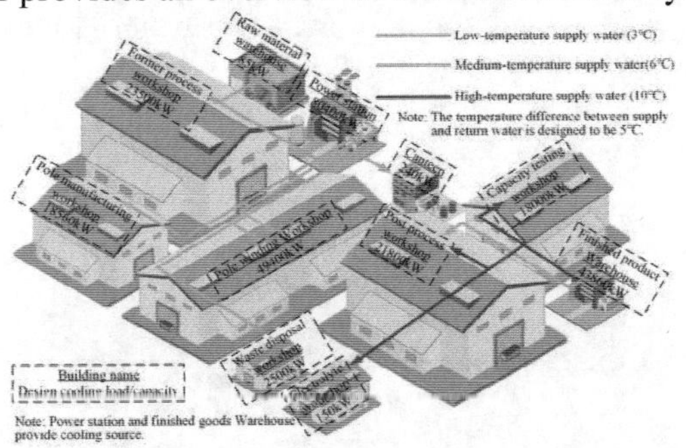

Figure 1. Axonometric drawing of the chilled water system.

2.2. The production process and the corresponding environmental requirement

The overall production can be divided into three processes: pole manufacturing, cell synthesizing, and capacity testing. Ternary lithium battery need to strictly control the humidity in the production environment. The specific process and humidity requirements are shown in **Figure 2**.

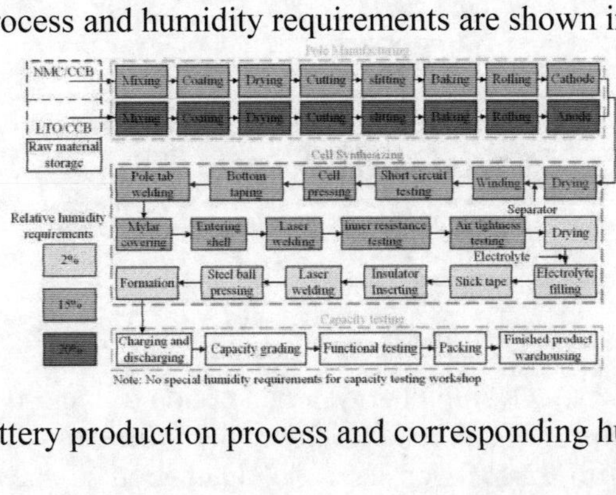

Figure 2. Lithium battery production process and corresponding humidity requirements.

2.2.1. Pole Manufacturing
The raw and auxiliary materials used in the project are all purchased. In order to avoid mutual contamination of cathode and anode materials, the raw material storage and the pole manufacture are separated into individual room. The main processes are: mixing, coating, rolling and slitting.

2.2.2. Cell Synthesizing
It is essentially an assembly process to complete the manufacturing of the cell, specifically the orderly assembly of the poles, with the separator and electrolyte. The main processes are: winding, electrolyte filling, welding, etc.

2.2.3. Capacity testing
After the completion of each process, the battery cells are tested for capacity and *K* value (voltage drop per unit time). This process can activate the battery to form a finished product that is safe to use and stable in performance. The main processes are: capacity grading, testing and so on.

3. Energy consumption analysis

3.1. Overall energy use status
Production process energy consumption of electricity, natural gas and water. Electricity consumption is divided into production electricity used directly by each process machines and facility electricity consumed by facility systems; natural gas is used for steam boiler to produce regenerative heat source for dehumidification rotor and heat transfer oil boiler to provide heat source for coating process; water is used to produce pure water and provide cooling make-up water. **Figure 3(a)(b)** shows the overall energy use of the project from August 2022 to July 2023.

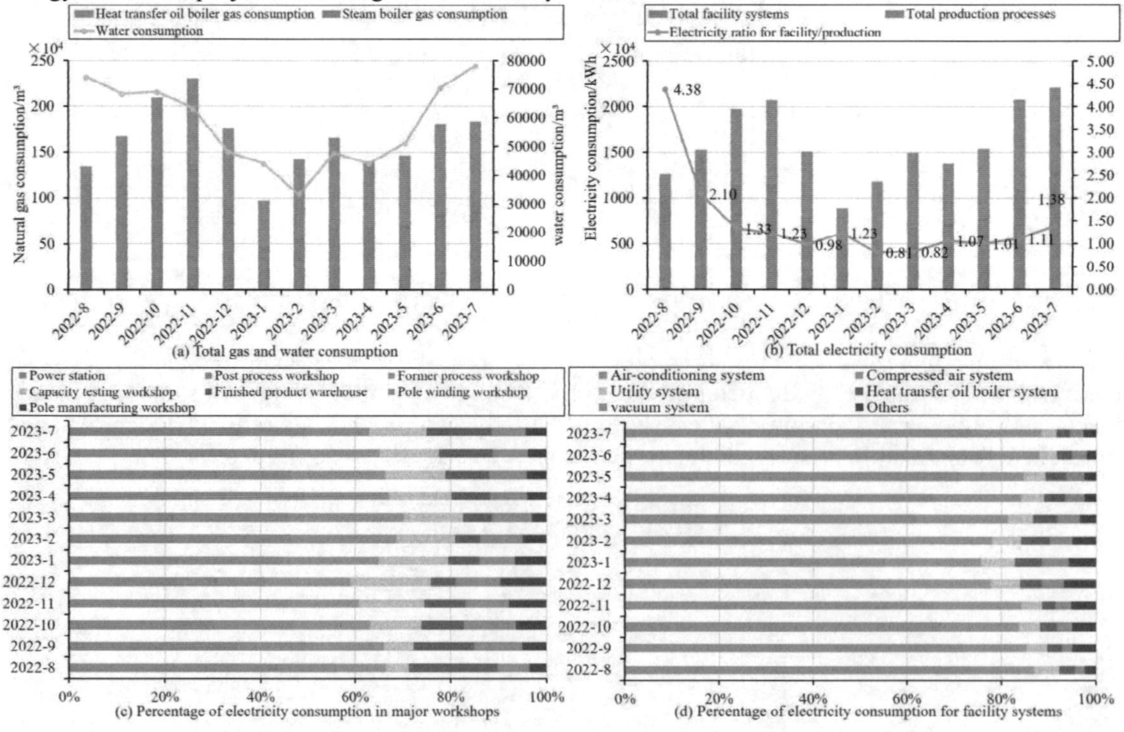

Figure 3. Overall energy consumption of the project.

Monthly natural gas consumption ranges from 0.95Mm³ to 2.3Mm³, monthly electricity consumption from 9.2MkWh to 22.1MkWh, and the fluctuation of the trend of gas and electricity consumption is more synchronized, both in November and July to achieve maximum value, while

January achieve minimum value due to the holiday shutdown. Monthly water consumption is between 33000m³~79000m³. Overall facility electricity use is higher than production process, and facility electricity use corresponding to unit production electricity use shows high in winter and low in summer. As in **Figure 3(c)**, The power station and finished product warehouse produce chilled water, so the energy consumption ratio is high in summer and low in winter. The capacity testing workshop has an energy consumption ratio higher than 10% except for the commissioning months. The ratio and fluctuation of energy consumption in the former and post process workshop are similar, ranging from 10% to 25%. The ratio of the pole winding and pole manufacturing is lower than 10% monthly.

3.2. *Facility Electricity Consumption Split*

The five main energy-consuming facility systems are: air-conditioning, compressed air, thermal oil boiler, vacuum, and utility. **Figure 3(d)** illustrates the percentage of electricity.

Air-conditioning system as the main facility energy-consuming system, monthly accounted for more than 55% of the total electricity consumption, in the winter due to the cooling load and cooling water temperature is low, accounting for a relatively low percentage of 55% to 60%, while in May to September accounted for more than 70%; compressed air system to provide air source for automated production lines, and produce nitrogen, as the second largest energy-consuming facility system accounted for 10% to 21%; the utility, heat transfer oil boiler, the vacuum accounted for less than 10% each month.

3.3. *Energy consumption of air-conditioning system*

The air-conditioning system can be categorized into cooling source and terminal air-conditioning unit. Except for the initial two commissioning months, the energy consumption of terminal unit is strongly correlated with the production consumption, the ratio of electricity consumption is basically stable at 0.2~0.3; the energy consumption of cooling source is affected by the cooling load and the outside temperature, and the ratio is more than 0.6 from April to November, and less than 0.4 from December to March. The air-conditioning system consumption accounts for up to 45% of the overall total electricity consumption of the project (June 2023). Therefore, the energy saving of cooling source system can be a key point to enhance the competitiveness of the product.

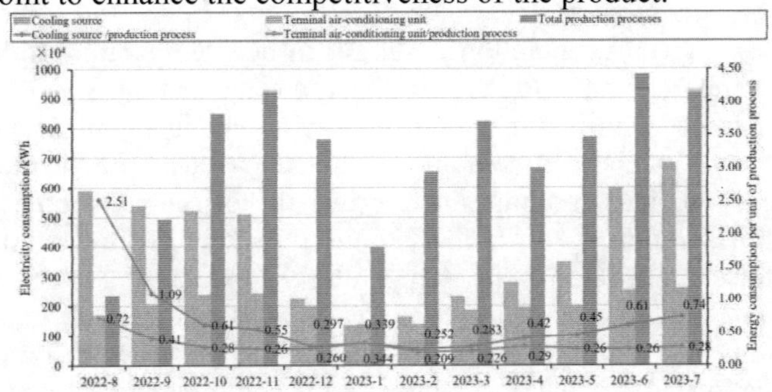

Figure 4. Electricity consumption relationship between air-conditioning system and production.

4. Research on the application of cooling tower free cooling technology

This section introduces the winter operating situation of air-conditioning system, theoretically analyzes the chilled-water-use characteristics of terminal equipment, optimizes the chilled water supply temperature, and explores energy-saving technology based on the optimized temperature.

4.1. *Analysis of the current situation of winter cooling and equipment performance*

The main supply terminal equipment for chilled water in the power station includes rotary dehumidifier (DHU double rotor/ SDHU single rotor), make-up air unit MAU, fan coiler FCU, process cooling water PCW, and air-handling unit AHU.

Chilled water is supplied throughout the day for production needs. In summer, the chilled water system is divided into low-temperature (3°C supply) and medium-temperature (6°C supply), and the low-temperature water is only supplied to the front and rear surface coolers of the DHU/SDHU in the winding workshop. In winter, the medium- and low-temperature chilled water pipelines are combined. The chilled water temperature is 10°C when the humidity ratio of the fresh air is lower than 7g/kg(a); and the temperature is 6°C when the humidity ratio of the fresh air is higher than 7g/kg(a). Neither the FCU nor the MAU operates in winter, i.e., the chilled water is only for DHU and PCW in winter.

Firstly, select the winding workshop DHU-WP2-04 operation data for analysis, time granularity of 5 minutes, screening time period in which fresh air humidity ratio is lower than 7g/kg(a). So fresh air surface cooler only plays a role in cooling but not dehumidifying during this period. Meanwhile, the air supply dew temperature set point, the air valve and fan frequency have not been adjusted during this period. Spearman's rank correlation test is performed on the parameters of fresh air humidity ratio and temperature, front rotor inlet air temperature and regeneration air temperature that may affect the supply air dew point and heat map is drawn as in **Figure 5**.

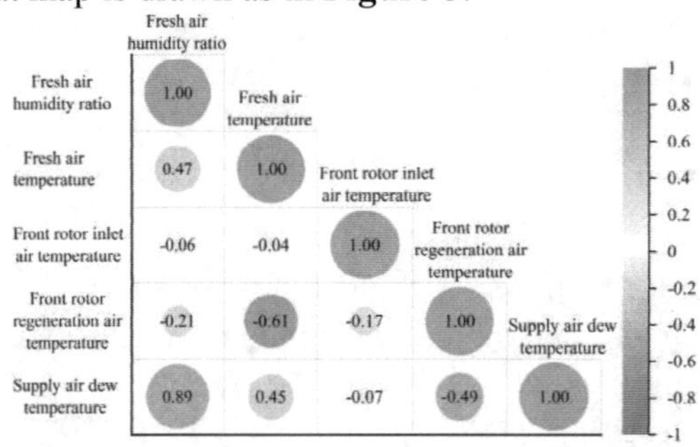

Figure 5. Correlation coefficients between winter operating parameters of DHU-WP2-04.

From **Figure 5** can be further deduced: when the fresh air humidity ratio below a certain value, the surface cooler only cools the new air, so that the lower rotor inlet air temperature strengthens the cooling and heating offset at the rotor, the rotor regeneration temperature instead of a certain degree of increase (weak negative correlation between rotor inlet air temperature and front rotor regeneration air temperature), but there is not a clear reduction in supply air dew temperature (inlet air temperature and the supply air dew temperature is not correlated). This indicates that the lower humidity ratio of fresh air in winter, to the extent that supply air dew temperature can be met, chilled water should be as high as possible to avoid the rotor inlet air temperature is too low to exacerbate the offset phenomenon.

The use of cooling tower free cooling can improve the offset phenomenon and reduce energy consumption of the refrigeration system, that is, in the case of low fresh air humidity ratio, dehumidifier demand for chilled water supply temperature is higher, and at this time the low fresh air humidity results in lower outlet water from cooling tower, which can be matched with the higher chilled water temperature. Therefore, the use of cooling tower in winter not only saves energy consumption of refrigeration system, but also alleviates the cooling and heating offset.

4.1.1. Cooling tower performance analysis

Figure 6 shows the thermal characteristic curve of the cooling tower. In order to make the same outlet temperature, a lower flow rate of inlet water can be realized at a higher wet bulb temperature; and in the case of the same amount flow rate of the inlet water, the smaller the cooling water temperature difference corresponds to a higher wet bulb temperature. Therefore, the application of smaller inlet flow rate and cooling water temperature differentials allows the cooling tower to operate at higher wet bulb temperatures, extending the free cooling hours[11].

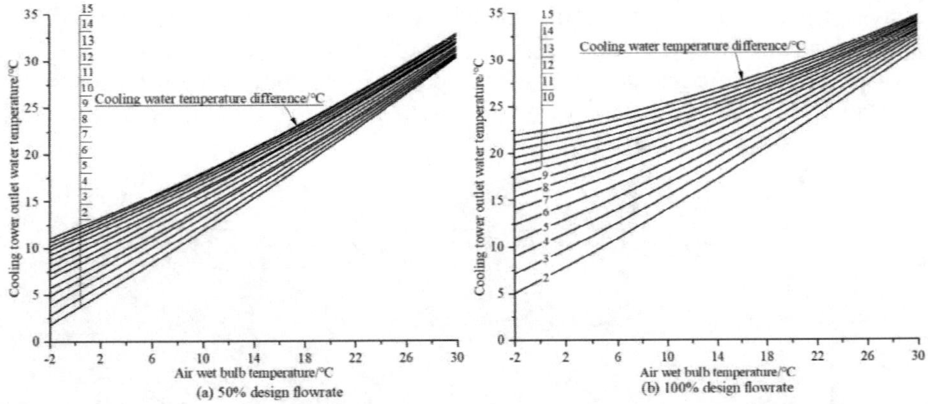

Figure 6. Thermal characteristic curve of cross-flow cooling tower.

4.1.2. PCW Performance analysis

PCW is set up in the former process and pole manufacturing workshop for rolling process cooling and solvent recovery. The secondary side water supply temperature is controlled by the opening of the primary side return water regulating valve. PCW parameters are shown in **Table 1**. In order to meet the cooling demand, there must exist an upper temperature limit for the inlet chilled water, and the following uses the *ε-NTU* method[12] to calculate this temperature limitation.

Table 1. PCW Parameters.

Workshop	Number of heat exchangers	U_A/ W/°C	Secondary side design water temperature/°C	Secondary side design flowrate/m³/h	Maximum flowrate of chilled water/m³/h
Former process	2	192500	18/24	320	312
Pole manufacturing	2	157500	18/24	285	340

$$\varepsilon = \frac{1-\exp\left[-\frac{UA}{C_{\min}}\left(1-\frac{C_{\min}}{C_{\max}}\right)\right]}{1-\left(\frac{C_{\min}}{C_{\max}}\right)\exp\left[-\frac{UA}{C_{\min}}\left(1-\frac{C_{\min}}{C_{\max}}\right)\right]} \quad (1)$$

From **equation (1)**, the former process and pole manufacturing heat exchanger(HX) effectiveness ε are 0.52 and 0.51, respectively, then the upper limit of the chilled water temperature can be obtained from **equation (2)**.

$$T_{c,in} = T_{h,in} - C_h\left(T_{h,in} - T_{h,out}\right)\left(\varepsilon C_{\min}\right)^{-1} \quad (2)$$

The upper limit chilled water temperature for the PCW are 12.1°C and 12.2°C of the former process workshop and the pole manufacturing workshop, respectively, resulting in the upper chilled water temperature limit for PCW being 12.1°C.

4.1.3. Rotary dehumidifier performance analysis

Correlation analysis shows that, under the premise of meeting the supply air humidity requirement, to increase the rotor inlet air temperature is energy-saving. Dehumidifier regeneration temperature need to set the upper and lower limits, and front and rear rotor regeneration temperature upper limit is 80°C in winter. The latter analysis is based on these two premises.

Dehumidifier is divided into fresh-air, mixed-air and return-air type. Fresh-air type DHU dehumidification load is the largest in the same amount of supply air volume. Due to the lower load in winter, fresh-air type DHU individually carries the room load, so only need to count in the fresh-air type DHU. From 2022.12.27 to 2023.1.17 supply air daily average dew temperature of each room fresh-air type DHU is shown in **Figure 7**.

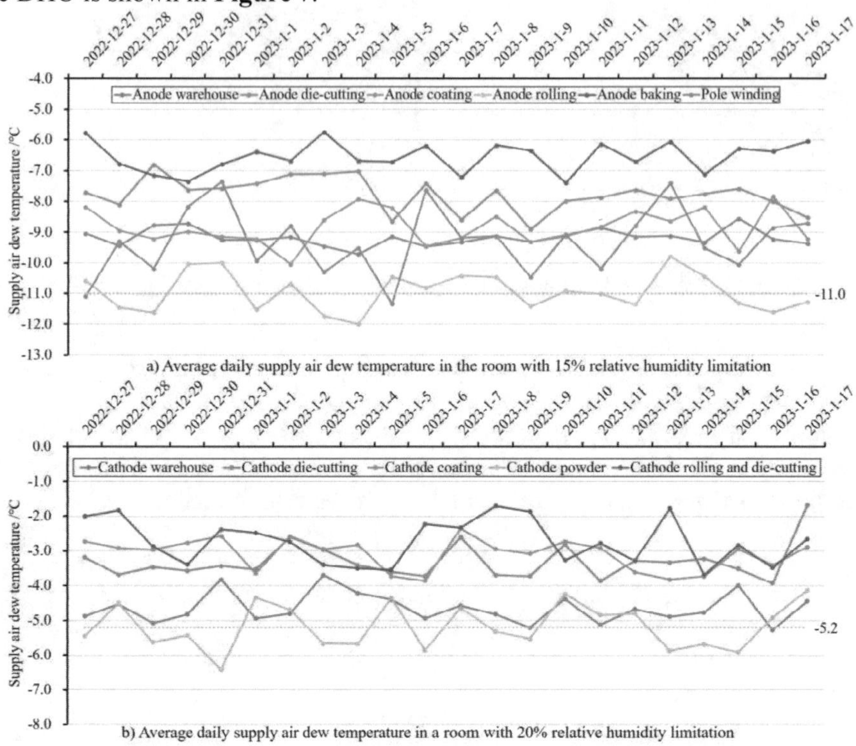

Figure 7. Supply air dew temperature for different rooms.

It can be seen that the anode rolling room demand for supply air dew temperature is the lowest, -11°C, corresponding humidity ratio is 1.46g/kg(a), that the dehumidifier dehumidification demand is considered to be the highest, that is, the chilled water demand temperature is the lowest, **Figure 8** demonstrates the corresponding air handling flow chart.

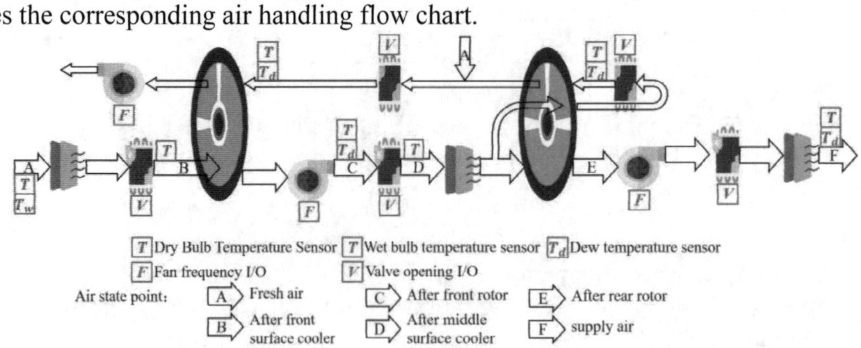

Figure 8. DHU-GY-01 air handling flow chart.

Jurinak[13] developed isopotential line model for silicone rotors to analyze dehumidification characteristics. The air state points are solved using isopotential lines F1 and F2 as in **equation (3)**.

$$\begin{cases} F1 = -2865T^{-1.49} + 4.344\omega^{0.8624} \\ F2 = \dfrac{T^{1.49}}{6360} - 1.127\omega^{0.07969} \end{cases} \tag{3}$$

Where T is the temperature in K; ω is the humidity ratio in kg(H_2O)/kg(a).

Schultz[14] proposed two effectiveness values $\varepsilon_1/\varepsilon_2$ to correcting model for non-ideal cases.

$$\begin{cases}\varepsilon_{F1}=(F1_D-F1_P)(F1_R-F1_P)^{-1}\\ \varepsilon_{F2}=(F2_D-F2_P)(F2_R-F2_P)^{-1}\end{cases} \tag{4}$$

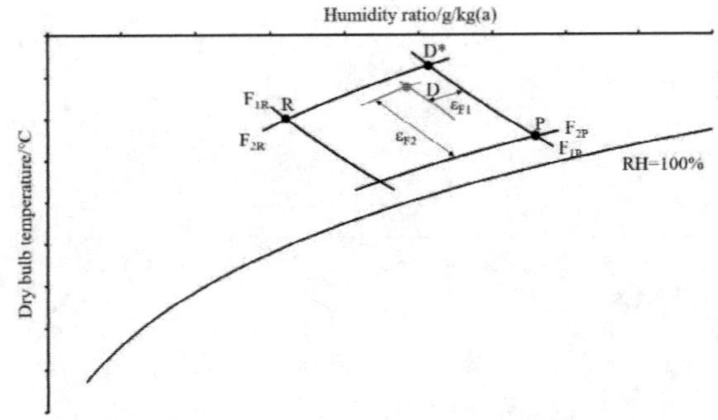

Figure 9. Isopotential line model schematic.

$$\begin{cases}F1_D=F1_{P,now}+\varepsilon_{F1,des}\left(F1_{R,now}-F1_{P,now}\right)\\ F2_D=F2_{P,now}+\varepsilon_{F2,des}\left(F2_{R,now}-F2_{P,now}\right)\end{cases} \tag{5}$$

In **Table 2**, actual operating conditions were selected to verify the model accuracy, and all the selected conditions were maintained for at least 30 minutes.

Table 2. Isopotential model error analysis.

Inlet air parameters				Outlet air parameters				Relative error/%	
Inlet air		Regeneration air		Model calculation		Actual			
ω /g/kg(a)	T/°C	T/°C	ω /kg/kg(a)	T/°C	ω /kg/kg(a)	T/°C	ω /kg/kg(a)	T	ω
6.2	10.1	80.1	13.5	24.2	2.2	25.2	2.3	-4.0	-4.3
6.0	11.9	78.5	13.0	25.2	2.3	26.2	2.2	-3.8	4.5
5.5	10.0	70.2	13.0	21.4	2.4	23	2.3	-7.0	4.3
5.0	6.1	65.3	12.6	16.5	2.2	17.8	2.1	-7.3	4.8
4.3	6.0	60.1	12.2	14.7	2.1	16.1	2.0	-8.7	5.0

The errors of T and are all controlled within 10% to support the subsequent calculations. The rotor dehumidification characteristics are derived using the isopotential line model as shown in **Figure 10**.

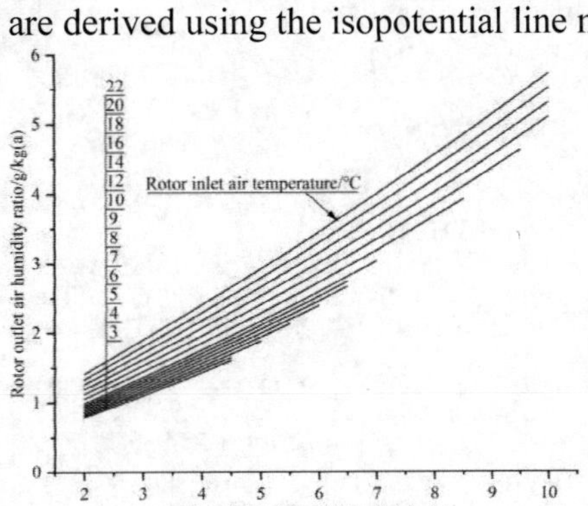

Figure 10. Characteristic curve of rotary dehumidification.

The dehumidifier front and rear rotor is the same, then the required supply air humidity can be deduced from the fresh air humidity. The rotor inlet air temperature corresponding to each humidity ratio can be obtained by making a reference line in **Figure 10**, as shown in **Table 3**.

Table 3. Upper limit inlet air temperature for humidity ratio.

ω /g/kg(a)	Optimized rotor inlet air temperature/°C
7.5	10.1
7.0	10.8
6.5	12.3
6.0	14.0
5.5	15.1
5.0	16.3
4.5	18.0
4	19.9
3.5	21.7

Surface cooler heat transfer also satisfies the *ε-NTU* method. The defining equation for dry working condition heat transfer is as follows.

$$Q_{dry} = \varepsilon_{dry} m_a C_{pm} \left(T_{a,i} - T_{w,i}\right) \tag{6}$$

The equation for ε_{dry} is

$$\varepsilon_{dry} = \frac{1-\exp\left(-NTU_{dry}\left(1-C^*\right)\right)}{1-C^*\exp\left(-NTU_{dry}\left(1-C^*\right)\right)} \tag{7}$$

Where,

$$\begin{cases} C^* = m_a C_{pm}\left(m_w C_{pw}\right)^{-1} \\ NTU_{dry} = UA_{dry}\left(m_a C_{pm}\right)^{-1} \end{cases} \tag{8}$$

ε_{dry} is related to heat transfer coefficient U, the area A_{dry}, the specific heat of the air flowing C_{pa} and the specific heat of the water flowing C_{pw}. Therefore, the surface cooler in the fixed air/water flowrate, ε_{dry} is identical. According to the actual dry working condition of the surface cooler, ε_{dry}=0.96.

Surface cooler air temperature and water temperature to meet the **equation (9)**.

$$T_{a,o} = T_{a,i} - \varepsilon_{dry}(T_{a,i} - T_{w,i}) \tag{9}$$

In **Figure 11**, the optimized rotor inlet air temperature is the upper limit of the rotor inlet air temperature corresponding to a certain air humidity ratio, substituting into the **equation (9)** to obtain the required chilled water temperature under the theoretical highest fresh air temperature situation, i.e., the upper limit of the fresh air surface cooler water temperature, as shown in **Table 4**.

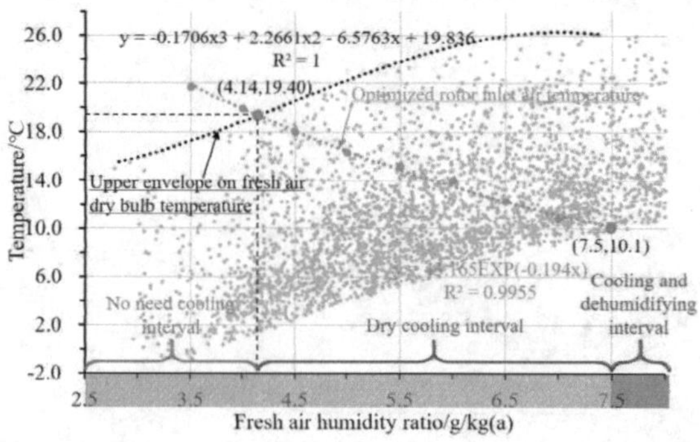

Figure 11. Outdoor air status and rotor inlet air temperature.

As shown in **Figure 11**, when the humidity ratio is lower than 4.14g/kg(a), the fresh air temperature is lower than the optimized rotor inlet air temperature, which indicates that the fresh air does not need to be cooled to reach supply air dew temperature. And at the point (7.5g/kg(a), 10.1°C), the air relative humidity has reached the apparatus dew point, indicating that when the fresh air humidity ratio is higher than 7.5g/kg(a), the surface cooler needs to dehumidify so that supply air humidity requirement 1.46g/kg(a) can be achieved. When the fresh air humidity ratio is between 7.5 and 4.14g/kg(a), the rotor inlet air should be cooled to the optimized inlet air temperature at least.

Table 4. Optimized rotor inlet air temperature and upper limit of fresh air surface cooler water temperature.

ω / g/kg(a)	Optimized rotor inlet air temperature/°C	Theoretical highest fresh air temperature/°C	Upper limit of fresh air surface cooler water temperature/°C
7.5	10.1	26	9.5
7.0	10.8	26.1	10.2
6.5	12.3	26	11.8
6.0	14.0	24.5	12.8
5.5	15.1	23.5	14.1
5.0	16.3	22	15.6
4.5	18.0	21	18.7
4	19.9	19	No cooling required

Middle surface cooler needs to cool the air warmed up by the front rotor to the upper limit of the rotor inlet air temperature, and the effectiveness of the middle surface cooler ε_{dry}=0.87, and then obtain the upper limit of middle surface cooler chilled water temperature as **Table 5**.

Table 5. Upper limit of middle surface cooler water temperature.

ω / g/kg(a)	Highest front rotor outlet air temperature/°C	Upper limit of middle surface cooler water temperature/°C
7.5	25.0	7.8
7.0	25.1	8.8
6.5	25.0	10.4
6.0	26.7	12.1
5.5	27.1	13.3
5.0	27.5	14.6
4.5	28.4	16.5

In summary, there is a mapping relationship between the upper limit of the chilled water temperature and the fresh air humidity ratio, as shown in **Figure 12**. The fresh air humidity ratio is low so that the dehumidifier demand for chilled water temperature shows a certain rise.

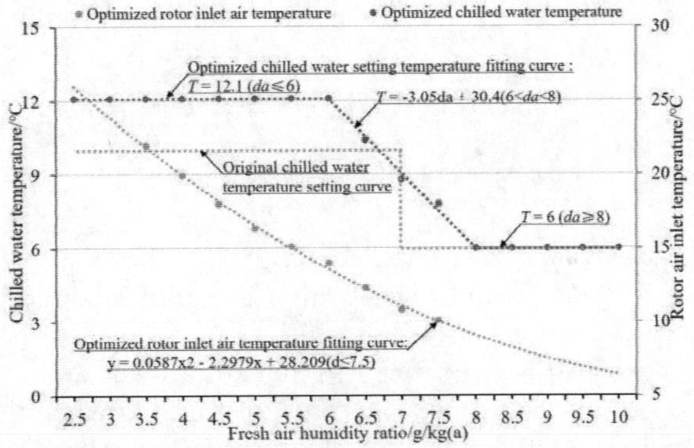

Figure 12. Optimized chilled water temperature and rotor inlet air temperature.

4.2. Control strategies and modeling of cooling tower free cooling technology

The main control strategy includes cooling mode switching control, number of chiller control, number of cooling tower control and cooling tower outlet temperature control. The low-temperature units need to be activated when high temperatures occur in the workshop, so the low-temperature refrigeration units are not involved in technology application. **Figure 13** shows the schematic, and the main equipment of the medium-temperature units is shown in **Table 6**.

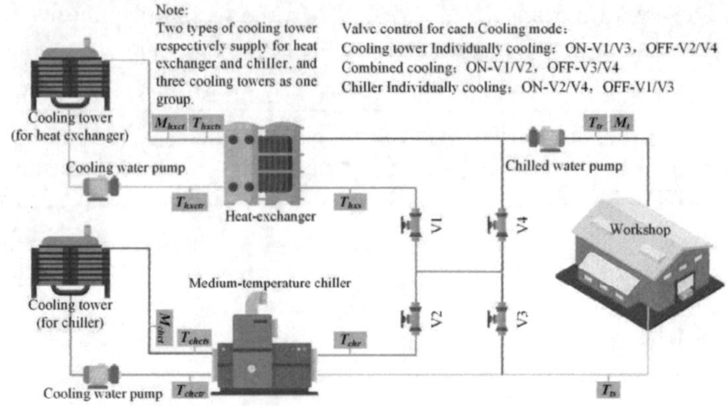

Figure 13. Cooling tower free cooling schematic.

Table 6. Main equipment of the medium-temperature unit parameters.

Equipment type and parameters	Rated input power/kW
Chiller, rated cooling capacity is 10200kW	1937
Chilled water pump, rated flowrate is 1930m³/h, head is 55.5m	400
Cooling water pump, rated flowrate is 2280m³/h, head is 25m	200
Cooling tower, rated flowrate is 820m³/h	30

4.2.1. Cooling mode switching control

There are three cooling modes: cooling tower individual cooling, combined cooling, and chiller individual cooling. According to the HX cooling tower outlet water temperature T_{hxcts}, terminal return water temperature T_{tr}, HX secondary side outlet water temperature T_{hxs} and chilled water supply temperature T_{ts} to determine the cooling mode. The temperature difference between the primary side inlet water and the secondary side outlet water of the heat exchange is taken as 1.5°C.

4.2.2. Number of chiller control

When cooling mode is cooling tower individual cooling the number of chiller in operation N_{ch}=0。When the cooling mode is combined cooling or chiller individual cooling, N_{ch} is calculated from **equation (10)**.

$$\begin{cases} Load_{ch}=M_t c_p \left(T_{hxs}-T_{ts}\right) \\ N_{ch}=INT\left(Load_{ch}\cdot Load_{Cap,ch}^{-1}\right)+1 \end{cases} \tag{10}$$

4.2.3. Number of cooling tower control

Cooling tower for chiller, as originally controlled, with a group of 3 cooling towers per chiller that is $N_{chct}=3N_{ch}$.

Cooling tower for HX need to limit the flowrate into the tower and the temperature difference between the water in and out, limit the former is to prevent freezing when the temperature is too low, this paper limits the cooling tower flowrate of 50% design flow rate; limit the latter because the smaller the cooling water temperature difference the longer the free cooling time, but it leads to an increase in the energy consumption of the cooling water pumps and a decrease in the overall energy efficiency. Adopting a cooling tower temperature difference of 2~3°C is energy efficient[15], therefore, the temperature difference ΔT_{hxct} taken as 3°C when calculating the required number of units. FLOOR function represents downward rounding to a multiple of 3, that is, to obtain the number of cooling tower enabled units, calculated as **equation (11)**.

In summary, the number of cooling tower controller will output $3N_{ch}+N_{hxct}$ when the cooling mode is cooling tower individual cooling or combined cooling.

$$\begin{cases} Load_{hxct} = M_t c_p \left(T_{tr} - T_{ts}\right) \\ N_{hxct} = FLOOR\left(\dfrac{Load_{hxct}}{50\% M_{Cap,ct} c_p \Delta T_{hxct}}, 3\right) \end{cases} \tag{11}$$

4.2.4. Cooling tower outlet temperature control

Cooling tower outlet water temperature control are realized by PID, by controlling the relative fan speed *RFS* so that the cooling tower outlet temperature meet the requirement.

The cooling tower for chiller, the lower the cooling water outlet temperature the higher the *COP* of the chiller, but in order to ensure a certain amount of cooling capacity, need to limit the condensing pressure, i.e., the cooling water outlet temperature can not be too low. Under the requirements of the chiller manufacturer, the chiller can be utilized as a minimum condenser water inlet of 18 °C.

The cooling tower for HX, following the principle of cooling tower priority cooling, should increase the fan speed as much as possible to maximize the use of natural cooling source. However, when the outdoor wet bulb temperature is very low and the cooling mode is cooling tower individual cooling, the water temperature controller needs to control the *RFS* according to the terminal demand.

In summary, each key control logic and TRNSYS simulation model are obtained as follows.

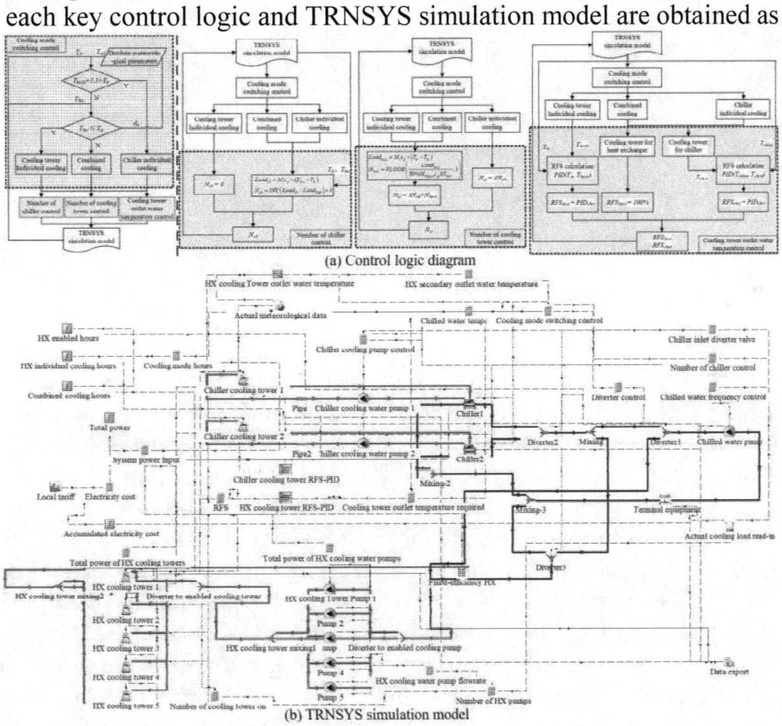

Figure 14. TRNSYS simulation model.

4.3. Analysis of energy-saving potential

This paper adopts the ratio of free cooling *RFC* to evaluate the energy saving effect of free cooling technology, which is more accurate compared to the free cooling hours.

$$RFC = \frac{Q_{FC}}{Q_t} \times 100\% \tag{12}$$

Where Q_{FC} is the cooling capacity provided by the cooling tower, Q_t is the total terminal cooling load.

Based on this index, this paper introduces the evaluation period, according to the DHU winter operating condition to determine the evaluation period from December 1 to March 15(the following year), the average *RFC* is obtained by integrating and averaging over time.

Firstly, adjust the cooling mode switching controller so that the cooling mode is always chiller individual cooling, and obtain the simulated energy consumption of the power station chilled water system in the evaluation period, as shown in **Table 7**.

Table 7. Simulated and actual energy consumption.

Electricity consumption	December	January	February	March(15 days)
Actual/kWh	1336441	664791	958758	562402
Simulated/kWh	1247499	690137	984773	574740
Relative error/%	6.7	-3.8	-2.7	-2.2

The overall energy consumption of the model deviates slightly from the actual, indicating that the simulation platform is able to accurately identify the system operating characteristics. Base on the TRNSYS model, simulate and compare the energy consumption difference of the free cooling technology before and after chilled water temperature optimization, as shown in **Figure 15**.

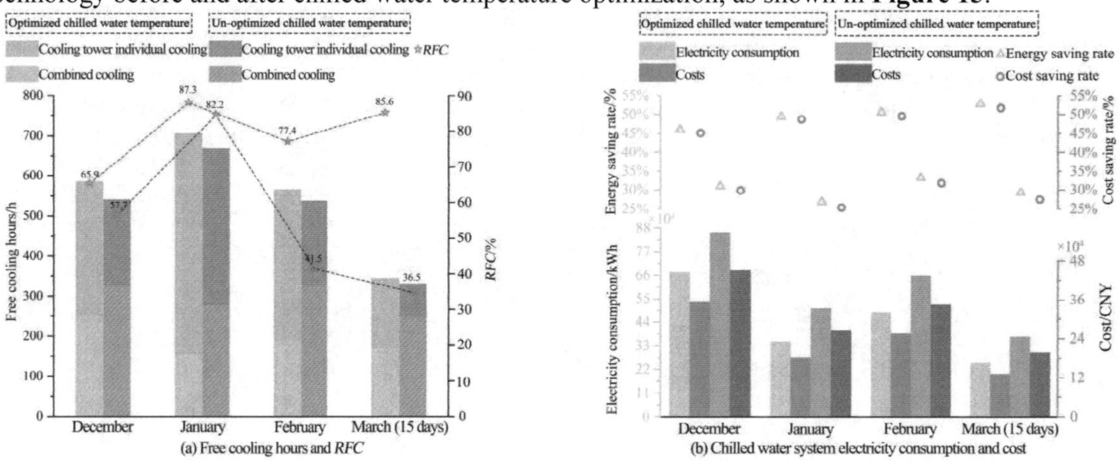

Figure 15. Analysis of the energy saving potential of free cooling technology.

The enhancement of cooling tower individual cooling hours is significantly stronger than the enhancement of overall free cooling hours, and the former have been extended by 40%~110%, which indicates that the optimized chilled water temperature has converted part of the combined cooling hours into individual cooling hours, resulting in the significant improvement of the *RFC*. The enhancement of individual cooling hours is the largest in March therefore the *RFC* enhancement rate was up to 49%, and optimizing the water supply temperature Optimization of water supply temperature achieves the expected effect. The lower temperature in January made the *RFC* reach the highest value of 89%, and the total free cooling hours reached 706h; in February, due to the long precipitation period, the fresh air humidity ratio was too high, which made the proportion of individual cooling reduced, and the *RFC* decreased; in the initial 15 days of March, the *RFC* was lower than that of January, but due to the higher terminal load, the energy saving rate and the cost saving rate reached the highest values of 52.8% and 51.6%. Compared with the system without free cooling, the technology can achieve energy saving rate and cost saving rate of 25%~34% in each month of winter, and after optimizing the chilled water temperature is the two saving rates are increased by more than 15%. Overall, the cooling tower free cooling technology with optimized chilled water temperature saved 49.0% of the refrigeration system energy consumption during the evaluation period, corresponding to an electricity saving of 1.689 million kWh and a cost saving of 854 thousand yuan.

5. Conclusion

This paper focuses on the winter chilled-water-use characteristics of lithium battery plant, optimizes the chilled water temperature according to the fresh air humidity ratio, and introduces the cooling tower free cooling technology to further exploit the energy-saving potential based on the optimized chilled water temperature. The main conclusions are as follows.

1) Lithium plant air-conditioning system is the largest proportion of energy-consuming facility systems, monthly electricity consumption accounted for more than 55% of the facility consumption, and accounting for the highest proportion of total electricity consumption up to 50%. The energy consumption of the cooling source is dominant, except in January, the other months are higher than the terminal air conditioning unit, which is even twice as high in the summer months, so the application of the cooling source energy-saving technologies for the manufacture cost is very critical.

2) In the lithium manufacture project application of free cooling technology should be necessary to optimize the chilled water temperature, in order to maximize the use of natural cooling source. Winter fresh air humidity ratio is lower, on the one hand, the DHU dehumidification load decreases, during when chilled water temperature should be appropriate upward adjustment; on the other hand, making the cooling tower outlet water temperature lower. Therefore, using the cooling tower as the cooling source in winter becomes practical and advantageous. According to the former analysis, when the humidity ratio is below 7.5g/kg(a), rotor dehumidifier dehumidification needs is reduce so that the chilled water temperature can be gradually adjusted upward, and the maximum value of 12.1°C is achieved in 6g/kg(a); when the humidity ratio below 6g/kg(a), due to the need to ensure that the PCW cooling needs, the chilled water temperature can not be further increased to maintain at 12.1°C. Different from the fixed chilled water temperature, according to the humidity ratio optimization of chilled water temperature, can increase the cooling tower individual cooling time 60%, and achieve the cooling system for higher than 45% of the energy-saving rate and cost-saving rate during winter condition.

References

[1] Capasso C and Veneri O 2014 Experimental analysis on the performance of lithium based batteries for road full electric and hybrid vehicles *Applied Energy* **136** 921-30

[2] Tang X, Zhou Y, Gao F and Lai X 2023 Joint estimation of state-of-charge and state-of-health for all cells in the battery pack using "leader-follower" strategy *eTransportation* **15** 100213

[3] Bai X, Yu T, Ren Z, Gong S, Yang R and Zhao C 2022 Key issues and emerging trends in sulfide all solid state lithium battery *Energy Storage Materials* **51** 527-49

[4] Research S P I o 2023 2023 White Paper on the Development of Global Lithium Battery Industry. (https://www.shifair.com/informationDetails/86748.html)

[5] Zaman W and Hatzell K B 2022 Processing and manufacturing of next generation lithium-based all solid-state batteries *Current Opinion in Solid State and Materials Science* **26** 101003

[6] Qi L, Wang Y, Kong L, Yi M, Song J, Hao D, Zhou X, Zhang Z and Yan J 2023 Manufacturing processes and recycling technology of automotive lithium-ion battery: A review *Journal of Energy Storage* **67** 107533

[7] Yuan C, Deng Y, Li T and Yang F 2017 Manufacturing energy analysis of lithium ion battery pack for electric vehicles *CIRP Annals* **66** 53-6

[8] Ahmed S, Nelson P A and Dees D W 2016 Study of a dry room in a battery manufacturing plant using a process model *Journal of Power Sources* **326** 490-7

[9] Ma Z, Liu X and Zhang T 2023 Experimental investigation and effectiveness analysis of a desiccant wheel dehumidification system with low air humidity *Applied Thermal Engineering* **226** 120279

[10] Guan B, Ma Z, Wang X, Liu X and Zhang T 2022 A novel air-conditioning system with cascading desiccant wheel and liquid desiccant dehumidifier for low-humidity industrial environments *Energy and Buildings* **274** 112455

[11] Amado E A, Schneider P S and Bresolin C S 2021 Free cooling potential for Brazilian data centers based on approach point methodology *International Journal of Refrigeration* **122** 171-80

[12] Kays W M and London A L 1964 *Compact Heat Exchangers, 2nd Edition* (New York: McGraw-Hill)

[13] Jurinak J J 1982 Open cycle solid desiccant cooling : component models and system simulations. (Madison: University of Wisconsin)
[14] Schultz K J, Beckman W A and Mitchell J W 1983 The Performance of Desiccant Dehumidifier Air-Conditioning Systems Using Cooled Dehumidifiers. (Madison: University of Wisconsin)
[15] Mi R, Bai X, Xu X and Ren F 2023 Energy performance evaluation in a data center with water-side free cooling *Energy and Buildings* **295** 113278

Study on the thermodynamic and economic performance of power generation cycle with small temperature difference

Ning Hao[1], Chuanliang Liu[1*], Dexuan Wang[1], Zhenya Li[1] and Wenjie Bian[1]
[1]Shanghai Power Equipment Research Institute Co. Ltd, Shanghai200240, China

*Corresponding author. Email: liuchuanliang@speri.com.cn

Abstract. The low thermal efficiency and low economic benefits of power generation systems under small temperature difference conditions have always hindered their practical operation. In order to explore effective ways to achieve performance optimization of thermodynamic cycle electrical power generating systems under small temperature difference conditions, this study constructs a static analysis model of an organic Rankine cycle with small temperature difference, selects six different commonly used refrigerants, and investigates the influence of key parameters on the thermodynamic and economic performance of the system using difference working fluids. The outcomes indicate that key parameters such as evaporation temperature and condensation temperature play a critical role in the thermodynamic and economic performance of the power generation system. For the working fluids, R227ea provides the highest net power output, while ammonia (R717) is the best performer in regard to system thermal efficiency and economic performance.

Nomenclature

A	heat transfer area, m^2	Δ	difference
C_p	specific heat capacity, kJ/kg	β	Corrugate Angle
D_h	hydraulic diameter, m		
f	Darcy friction factor	*Subscript*	
h	specific enthalpy, kJ/kg	1,2,3…	state points
K	pipe roughness, m	con	condenser
L	Length, m	dry	dry fluid
$\dot{m}$	mass flow rate, kg/s	dw	deep seawater (cold water)
P	power, kW	eva	evaporator
Q	heat transfer, kW	exp	expander
Re	Reynolds number	is	isentropic
T	temperature, °C	net	net

v	velocity, m/s	p	pump
w	specific net output power, kW/kg	pp	pinch point
k	heat transfer coefficient, W/(m^2•K)	sh	superheated
		sw	surface seawater (hot water)
Greek symbols		tot	total
ρ	density, kg/m^3	wet	wet
η	efficiency	wf	working fluid
γ	net power output per unit area, kW/m^2	LM	Logarithm
λ	heat conductivity coefficient	max	maximum
δ	thickness of metal wall	min	minimum
α	convective heat transfer coefficient		

1. Introduction

For a long time, converting temperature difference energy into electricity through thermodynamic cycles under small temperature difference conditions has been widely regarded as a power generation method with great application prospects and potential[1-2]. For example, the vast amount of ocean thermal energy stored in the vast oceans is a form of small temperature difference energy. At low latitudes, the temperature of the surface ocean water remains between 25-30°C all year round, while deep ocean water at a depth of 1000~1200m maintains a temperature of 4~6°C[3-4]. Ocean thermal energy is a massive amounts of heat power stored in the temperature difference between different depths of seawater. It has abundant reserves, is clean and sustainable, and is considered the most promising renewable energy source for deep-sea electricity supply.

The key to utilizing small temperature difference energy to stably and sustainably supply electricity is to master the technology of converting small temperature difference energy into electricity through thermodynamic cycle systems[5]. The organic Rankine cycle(ORC) is one of the broadest and most commercially viable power generation cycle systems under small temperature differentials[6]. It uses low-boiling organic substances as working fluids to achieve power generation under small temperature differences. Due to the relatively small temperature differences in the system, the selection of different working fluids often leads to a significant impact on the performance of the heat exchanger, expander and thermal cycle. Therefore, the selection of a suitable working fluid is very significant in enhancing the performance of the system.

Scholars from various countries have conducted a considerable amount of research on the optimization of working fluids in ORC under small temperature difference conditions[7-9]. For example, Rosard et al. [10] compared ammonia with 11 other working fluids including propane and the results show that the volume of carbon dioxide turbine is the smallest when the generating power is the same. In the case of the same turbine volume, the power of the turbine with ammonia as the working fluid is twice that of the carbon dioxide turbine. Stoecher et al. [11] compared R134a,ammonia and R22 as working fluids, and the results certificate that under the same heat flow conditions, systems using ammonia as the working fluid cost up to 10 times less than R134a and R22. Kim et al. [12] developed computer modeling programs for five forms of thermal cycles, including the ORC, and compared several different working fluids. The outcomes suggest that the capabilities of R410a and R32 is comparable to that of R22 and ammonia. Sun et al. [13] optimized ammonia and R134a as working fluids with the objective of power output maximization, and the outcomes indicate that ammonia and R134a showed good performance in terms of thermal and exergy efficiencies. Yang et al. [14] quantitatively studied systems based on ORC and compared five commonly used working fluids: ammonia,

R600a, R245fa, R152a, and R134a, based on net output per unit area. The outcomes indicated that ammonia performed the best in the evaluation of target parameters, while R600a performed better in the evaluation of thermal efficiency. Yoon et al. [15] compared thermal efficiency and major component sizes of three working fluids, and the outcomes indicated that ammonia is the preferred fluid. However, ammonia can present environmental and safety hazards.

However, in the work of the above scholars, the index parameters to measure the system performance often fail to combine the utilization rate of heat source and the initial investment cost of system equipment. Based on it, this study aims to explore effective ways to improve the performance of thermodynamic cycle power generation systems under small temperature difference conditions. Firstly, based on the operating conditions of small temperature differences, common organic working fluids were selected, and six working fluids capable of small temperature difference ORC. Then, a static analysis model of ORC power generation system was constructed, and the rationality of the model was verified using literature data. Finally, through the constructed model, the effects of the key parameters of the system, such as evaporation temperature and condensation temperature, as well as different work fluids on the thermodynamic and economic performance of the system are investigated.

2. System description and model

The ORCs refers to the Rankine power cycle that uses organic refrigerants as working fluids. It is the most widely used and structurally simple cycle system in thermodynamic cycle power generation systems under small temperature difference conditions. This study takes the ORC power generation system based on ocean thermal energy as an example to investigate capacity of the ORC system for small temperature differences. The system diagram is illustrated as Figure 1. The main equipment involved are an evaporator, an expander, a condenser and a pump in the cycle. The liquid working fluid gain the heat and evaporated by the heat source in the evaporator. The evaporated vaporized working fluid flows into the expander to expand and generate electricity. The vapor from the expander flows into the condenser, where cold source water is liquefied. Finally, the liquefied working fluid is pumped into the evaporator and a novel cycle begins. In addition, due to the long length and large flux of the seawater pipeline, the power consumption of the seawater pump is to be also taken into account for modelling the system.

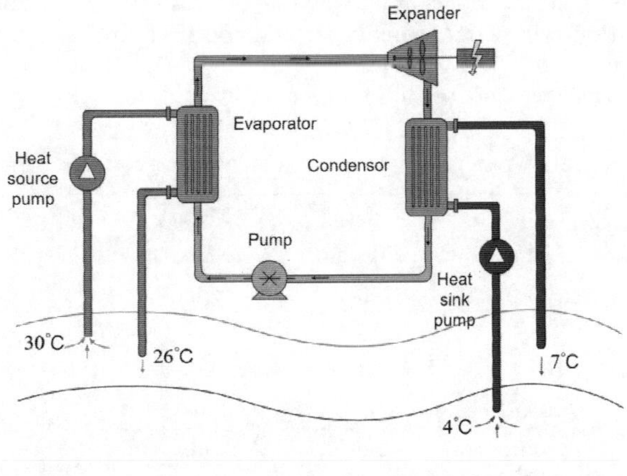

Figure 1. Schematic diagram of ORC power generation system using ocean thermal energy.

2.1. *Primary selection of working fluids*

The working fluid usually requires to satisfy the following requirements [16-19]:

(1) The working fluid should meet safety requirements (non-toxic or low toxicity, non-flammable, non-explosive, non-corrosive to pipelines and equipment).

(2) The working fluid should be environmentally friendly, that is, it should not cause damage to the ozone layer, and the Ozone Depletion Potential (ODP) should be 0.
(3) Chemical stability is necessary for the working fluid and its decomposition temperature significantly higher than the system operating temperature.
(4) Critical temperature should optimally be higher than the heat source temperature.
(5) It should be cost-effective and easily accessible.

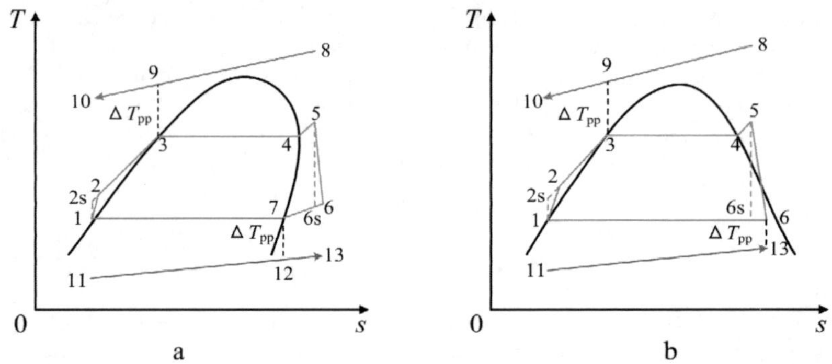

Figure 2. T-s Diagram (a. Dry fluid; b. Wet fluid).

Furthermore, drop of pressure in the piping reduces the pressure of the steam entering the expander, thus influencing the thermal efficiency of the system. Consequently, the pressure difference between the high and low pressures in the system loop should not be too small. In this study, with a reference evaporation temperature of 24°C and condensation temperature of 8°C, it is required that the pressure drop between the evaporator and the condenser is greater than 150 kPa. Under the above requirements, this paper considers six commonly used refrigerants with an ODP of 0 as alternative working fluids, consisted by three types of dry fluids and three wet fluids. The thermophysical properties of the six components are illustrated in Table 1. Dry fluids and wet fluids are divided on the basis of the dew point curve slope in the T-s diagram. A fluid whose dew point curve has a positive slope is a dry fluid, and a fluid whose slope is negative is a wet fluid, as shown in Figure 2. Generally, Dry fluids tend to superheat after the expander, and wet fluids are in a wet state.

Table 1. Thermophysical and environmental properties of alternative fluids.

Dry Fluid	Critical Temperature (°C)	Critical Pressure (Mpa)	Molar Mass (g/mol)	ODP	Safety Class
R1234yf	94.7	3.38	114.04	0	A2L
R1234ze(E)	109.4	3.64	114.04	0	A2L
R227ea	101.8	2.93	170.03	0	A1
Wet Fluid	Critical Temperature (°C)	Critical Pressure (Mpa)	Molar Mass (g/mol)	ODP	Safety Class
R152a	113.3	4.52	66.05	0	A2
R32	78.1	5.78	52.02	0	A2L
Ammonia(R717)	132.3	11.33	17	0	B2

2.2. Model Description

Based energy conservation principle, this paper develops the static model on the key facilities, including the evaporator, the expander, the compensator, the working fluid pump and cold water pump, and solves the governing equations of the static model. To simplify the calculations, certain assumptions need to be made for the system model. This model assumes that the power output of the expander is fixed at 50 kW. The power consumption of the pumps is calculated based on the pressure drop in the pipeline. In addition, made some general assumptions:

(1) System is in steady state operation.
(2) Heat exchange between the pipes, equipment, and the external environment is ignored.
(3) Pressure losses of the pipelines and heat exchangers are neglected.
(4) The flow velocity in the pipes remains constant.
(5) The physical properties of the hot and cold water are constant and incompressible.

The governing equations are listed as follows:

For the evaporator

$$Q_{\text{eva}} = \dot{m}_{\text{wf}}\left(h_5 - h_2\right) = \dot{m}_{\text{sw}} c_{\text{p}}\left(T_8 - T_{10}\right) \tag{1}$$

For a pure substance, the pinch point is in the saturated liquid phase, and its temperature difference is calculated by

$$\Delta T_{\text{pp}} = T_9 - T_3 \tag{2}$$

The evaporator outlet superheat is defined by

$$\Delta T_{\text{sh}} = T_5 - T_4 \tag{3}$$

The expander is a component that outputs power to the outside, and its power output is calculated by

$$P_{\text{exp}} = \dot{m}_{\text{wf}}\left(h_5 - h_6\right) \tag{4}$$

The isentropic efficiency of the expander is assumed to be constant and is calculated by

$$\eta_{\text{is,exp}} = \frac{h_5 - h_6}{h_5 - h_{6\text{s}}} \tag{5}$$

The energy conservation equation for the condenser is

$$Q_{\text{con}} = \dot{m}_{\text{wf}}\left(h_6 - h_1\right) = \dot{m}_{\text{dw}} c_{\text{p}}\left(T_{13} - T_{11}\right) \tag{6}$$

For a pure working fluid condenser, the pinch point appears at the saturated vapor phase, and its temperature difference is defined as

$$\begin{aligned} \Delta T_{\text{pp,dry}} &= T_7 - T_{12} \\ \Delta T_{\text{pp,wet}} &= T_6 - T_{13} \end{aligned} \tag{7}$$

The power consumption of the working fluid pump is defined as

$$P_{\text{p}} = \dot{m}_{\text{wf}}\left(h_2 - h_1\right) \tag{8}$$

The isentropic efficiency of the working fluid pump is presumed to be constant and is calculated by the rate of the isentropic enthalpy increase to the practical enthalpy increase

$$\eta_{\mathrm{is,exp}} = \frac{h_{2s} - h_1}{h_2 - h_1} \tag{9}$$

The power consumption of the cold water pump is calculated based on the pressure drop in the cold water pipeline

$$\Delta P_{\mathrm{sw(dw)}} = f \frac{\rho_{\mathrm{sw(dw)}} L_{\mathrm{sw(dw)}} v_{\mathrm{sw(dw)}}^{2}}{2D_{\mathrm{h}}} \tag{10}$$

where f is the Darcy friction factor, which is a dimensionless parameter calculated by

$$\frac{1}{\sqrt{f}} = -2\log\left(\frac{2.51}{Re\sqrt{f}} + \frac{K / D_{\mathrm{h}}}{3.72}\right) \tag{11}$$

The net power output of the system is defined as

$$P_{\mathrm{net,OTEC}} = P_{\mathrm{exp}} - P_{\mathrm{p}} - P_{\mathrm{sw}} - P_{\mathrm{dw}} \tag{12}$$

The net efficiency of the system is defined as

$$\eta_{\mathrm{OTEC}} = \frac{P_{\mathrm{net,OTEC}}}{Q_{\mathrm{eva}}} \tag{13}$$

The specific net output power of the system is defined as:

$$w_{\mathrm{net}} = \frac{P_{\mathrm{net,OTEC}}}{m_{\mathrm{sw}}} \tag{14}$$

In addition, a simple indicator is used to characterize the economy of the system, which is the net power output per unit area of heat transfer

$$\gamma = \frac{P_{\mathrm{net,OTEC}}}{A_{\mathrm{tot}}} \tag{15}$$

Where A_{tot} is the complete area of heat transfer of the system, and is calculated by

$$A_{\mathrm{tot}} = A_{\mathrm{eva}} + A_{\mathrm{con}} \tag{16}$$

The evaporator and condenser in the system are plate heat exchangers. According to the working fluid state, the evaporator is divided into three parts: liquid phase, two-phase and gas phase. For wet fluids, the condenser has only one part of the two-phase region, and for dry fluids, the condenser is divided into two parts: the gas phase and the liquid phase. The heat transfer area of each part can be calculated by Newton's cooling formula

$$A_{\mathrm{i}} = \frac{Q_{\mathrm{i}}}{k_{\mathrm{i}} \Delta T_{\mathrm{LM,i}}} \tag{17}$$

Logarithmic mean temperature difference is expressed as

$$\Delta T_{\mathrm{LM,i}} = \frac{\Delta T_{\max} - \Delta T_{\min}}{\ln\left(\frac{\Delta T_{\max}}{\Delta T_{\min}}\right)} \tag{18}$$

The total heat transfer coefficient is expressed as the sum of the thermal resistance of each heat transfer link

$$\frac{1}{k} = \frac{1}{\frac{1}{\alpha_1} + \frac{\delta}{\lambda} + \frac{1}{\alpha_2}} \tag{19}$$

The convective heat transfer coefficients of the seawater side and the working fluid side in the single-phase zone (supercooled zone and superheated zone) were calculated using Muley correlation. The evaporative heat transfer coefficient on the working fluid side of the evaporation zone is calculated by the Amalfi correlation equation. The condensing heat transfer coefficient of the working fluid side of the condensing zone is calculated by Shah correlation equation.

2.3. Model Validation

In an attempt to test the correctness of the static analysis of the model of the system constructed in this paper, a comparison analysis was conducted with the calculated results in reference[20]. The parameter settings are shown in Table 2-2 and the test results are summarized in Table 2-3.

Table 2. Parameter Settings in Reference[20].

Parameter	Value
Hot Source Temperature	28°C
Cold Source Temperature	5°C
Temperature difference at the pinch point	2°C
Expander Efficiency	85%
Working Fluid Pump Efficiency	80%
Hot/Cold Source Water Pump Efficiency	80%
Hot Source Water Pipeline	200m
Cold Source Water Pipeline	1000m

From the verification results in Table 3, the maximum error of the model parameters is 3.6%, which comes from the system net efficiency and is within an acceptable range, indicating that the model is reasonable.

Table 3. Verification Results.

	Results in Reference [20]	Results in Current Model	Error (%)
Evaporation Temperature, °C	25.5	25.5	-
Condensation Temperature, °C	9.5	9.5	-
Hot Source Water Flow Rate, kg/s	462.76	462.76	-
Hot Source Water Temperature Drop, °C	0.54	0.532	1.48
Cold Source Water Flow Rate, kg/s	95.56	94.183	1.44
Cold Source Water Temperature Rise, °C	2.5	2.5	0
Expander Power Output, kW	44.66	44.324	0.75

Working Fluid Pump Power Consumption, kW	0.67	0.663	1
Net Output of the System, kW	33.20	34.09	2.76
Net Efficiency of the System, %	3.32	3.44	3.6

3. Impact of Key Parameters on System Performance

Based on the established static analysis model, the system net efficiency, specific net output power, and economy are considered as indicators, and the influence of evaporation temperature, condensation temperature on system performance indicators is analyzed.

3.1. Evaporation Temperature

The Carnot cycle thermal efficiency is the highest theoretical thermal efficiency achievable for all thermodynamic cycles, where the evaporation temperature is equal to the temperature of the heat supply and the condensation temperature is equal to the temperature of the cold supply. Thus, all other things being equal, a higher evaporation temperature means a higher system thermal efficiency, closer to a theoretical maximum thermal efficiency. In addition, the increased evaporation temperature leads to an increase in the flow rate of the heat source water and an increase in the power consumption of the heat source pump due to the limitation of the temperature difference between the pinch points, leading to a decrease in the net efficiency of the system. Considering these two factors, the impact of evaporation temperature on the net efficiency of the system is shown in Fig. 3. It can be seen from the figure that the net efficiency of the proposed system improves with increasing evaporation temperature. In the example of ammonia (R717), the net efficiency of the system increased from 2.07% to 4.06% as the evaporation temperature increased from 18°C to 27°C. This suggests that the system thermal efficiency is more sensitive to changes in evaporation temperature than the power dissipation of the seawater pumps. Therefore, if a high net system efficiency is required, the evaporation temperature should be as high as possible. In terms of the choice of working fluids, as can be seen in the figure that R1234ze performs better in terms of net efficiency in the dry fluid, while ammonia (R717) has the highest net efficiency in the wet fluid, followed by R152a, and R32 performs the worst, and this difference also grows as the evaporation temperature increases.

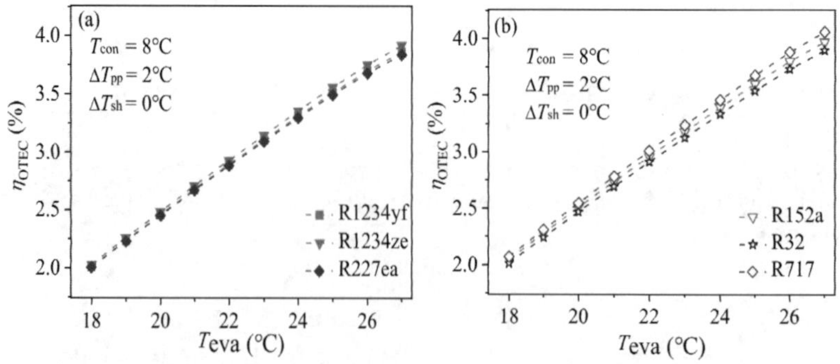

Figure 3. Impact of Evaporation Temperature on Net Efficiency of the System (a. Dry fluid; b. Wet fluid).

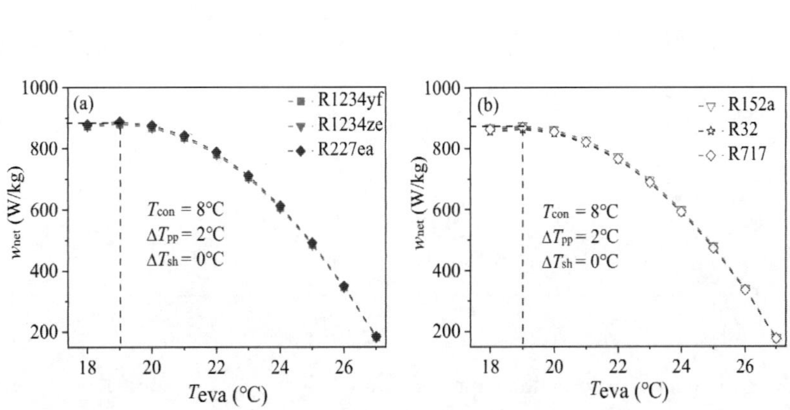

Figure 4. Impact of Evaporation Temperature on Specific Net Output Power of the System (a. Dry fluid; b. Wet fluid).

The specific net output power of the system is defined as the net output power of the system under the unit heat source water flow. This index can not only compare the output power capacity of each working fluid at the same level, but also take into account the utilization rate of heat source water. Figure 3-2 shows the effect of evaporation temperature on the specific net output power of the system. As can be seen from the figure, with the increase of evaporation temperature, the net efficiency of the system increases, but due to the limitation of the temperature difference of the evaporator pinch point, the seawater temperature at the evaporator outlet increases, and the water utilization rate of the heat source decreases. Therefore, the net specific output power of the system first increases and then decreases with the evaporation temperature, and reaches the maximum value around 19℃. Taking ammonia (R717) as an example, the net specific output power of the system reaches the maximum value of 870.53W/kg when the evaporation temperature is 19℃. In addition, no matter the dry fluid or the wet fluid, the specific net output power of each fluid is not much different, but the dry working fluid R227ea shows a slightly higher specific net output power than the other 5 working fluids.

The upfront capital expense depends on evaporator and condenser heat transfer area, especially if the heat transfer area is large. Thus, a simple metric was used to measure the economic feasibility, which is net power generation capacity of a unit area. Figure 5 illustrates the impact that evaporation temperature has upon system unit area in terms of clear power output. With increasing evaporation temperature, there is an increase in net efficiency as well as net system power output. However, higher evaporation temperatures result in a decrease of thermal heat exchange temperature difference (TTD), which requires a larger evaporator heat transfer area. Taking these two factors into account, net power output per unit area of the system initially rises with the evaporation temperature and then declines, peaking at 25°C. In addition, R227ea has the highest net power output per unit area in dry operating fluids, while in wet operating fluids, ammonia offers significant strengths compared to each of 2 other working fluids. At an evaporation temperature of 25°C, ammonia (R717) has a net per-area power output of 132 W/m^2 , while R32 has the greatest net per area power output among the other 5 working fluids, at 94.87 W/m^2 .

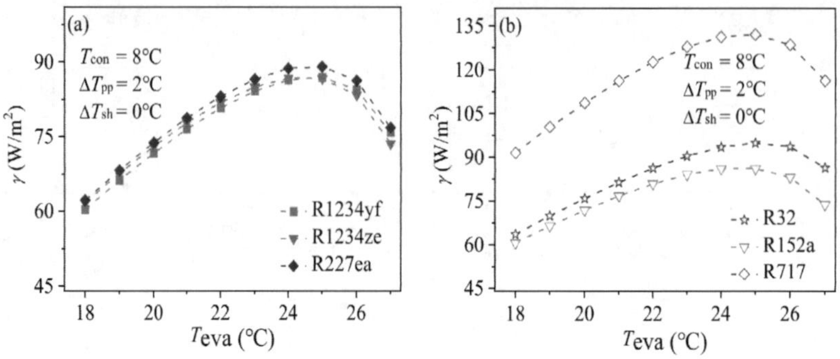

Figure 5. The impact of evaporation temperature on the net output per unit area of the system (a. dry fluid; b. wet fluid).

3.2 Condensation Temperature

In the Carnot cycle, a lower temperature of the cold source means a higher cycle thermal efficiency. Therefore, in the organic Rankine cycle, lowering the condensation temperature enhances the thermal effectiveness of the cycle. On the other hand, lowering the condensation temperature results in lower condenser outlet seawater temperatures, increasing the cold source water flow rate consumption and increasing the seawater pump consumption because of the limitation of the temperature difference between the pinch points. Effects of condensation temperature on this system are reflected in these two areas, as shown in Figure 6. As the figure indicates, the net efficiency of the system degrades monotonically as the condensation temperature is enhanced. In the case of ammonia (R717), for example, the net efficiency was reduced from 3.60% to 2.40% as the condensation temperature was increased from 8°C to 14°C.This indicates that the variation in seawater pump power consumption caused by the change in condensation temperature is not as significant as the impact on the system thermal efficiency.

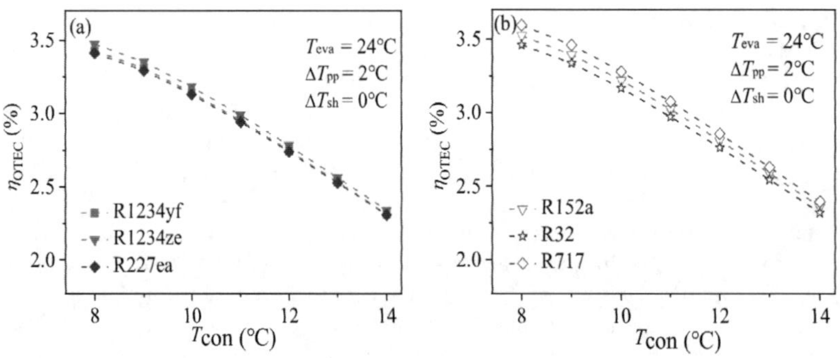

Figure 6. Influence of Condensation Temperature on System Net Efficiency (a. Dry Fluid; b. Wet Fluid).

As indicated in Figure 7, the specific net output monotonically reduced with the increase in condensation temperature of the system. In the case of ammonia (R717), for example, with a rise in condensation temperature from 8°C to 14°C, the net output reduces from 616.52W/kg to 401.81W/kg. Since the specific net output index only takes into account the utilization efficiency of the heat source, the influences from the condensation temperature are implemented by affecting the thermal efficiency and the power consumption of the seawater pumps. Accordingly, the variation of net output power with condensation temperature is in accordance with the variation in net system efficiency with condensation temperature.

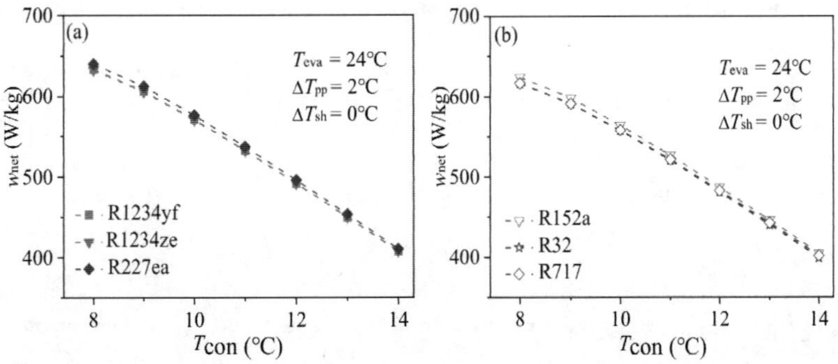

Figure 7. Influence of Condensation Temperature on System Specific Net Output Power (a. Dry Fluid; b. Wet Fluid).

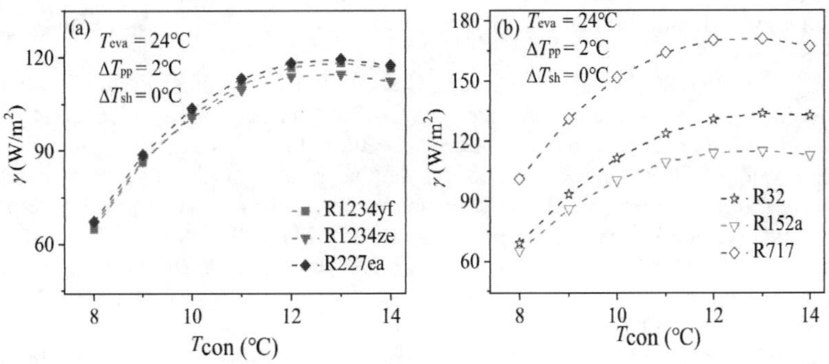

Figure 8. The Influence of Condensation Temperature on the Net Output Power per Unit Area of the System (a. Dry Fluid; b. Wet Fluid).

The effect of condensation temperature on the system output power per unit area is shown in Figure 8. The net efficiency of the system decreases with increasing condensation temperature and the net power output decreases per unit area. However, the heat transfer temperature difference within the condenser will also ıncrease, resultıng ın a smaller heat transfer area. Taking both of these factors into account, the net power output per unit area increases and then decreases, with a maximum at a condensation temperature of around 13°C. In the case of the R227ea, the net power output per unit area at a condensation temperature of 13°C is a maximum of 119.39 W/m^2.Ammonia (R717) exhibits the highest thermal efficiency and power output per unit area of any working medium, while R227ea has the highest specific net power output.

4. Conclusion

In this paper, by constructing a static analytical model of organic Rankine cycle power generation system under small temperature difference operating conditions and screening common organic workmasses, the key parameters of the system and the influence of different workmass fluids on the thermodynamic and economic efficiency of the system are studied, and the following findings are reached:

(1) The net efficiencies of the system increases monotonically with the rise in the evaporation temperature. The specific net output and net output per unit area initially increase and then decrease with increasing evaporation temperature. The maximal specific net output is achieved at an evaporation temperature of 19°C, and the maximal net output per unit area is achieved at a steam temperature of 25°C。

(2) Both net efficiency and specific net output decrease as the condensation temperature increases. As the condensation temperature increases, the net output per unit area shows a

tendency to increase and then decrease slightly, reaching a peak value at a condensation temperature of 13°C.

(3) Among dry working fluids, R227ea typically has the highest specific net output and net output per unit area, while the R1234ze has a higher net efficiency. And ammonia (R717) has significantly higher thermal and economic performance than other wet working fluids. Overall, dry and wet working fluids perform similarly in terms of thermal performance, but in terms of economic performance, ammonia (R717) outperforms all other working fluids.

Reference

[1] Garcia S I, Garcia R F, Carril J C and Garcia D I 2020 A review of thermodynamic cycles used in low temperature recovery systems over the last two years. *Renew. Sust. Energ. Rev.* **81** 760-767.

[2] Huo D X, Tian H, Shu G Q and Wang W G 2022 Progress and prospects for low-grade heat recovery electrochemical technologies. *Sustain. Energy Technol. Assess.* **49** 101802.

[3] Langer J, Quist J and Blok K 2020 Recent progress in the economics of ocean thermal energy conversion: Critical review and research agenda. *Renew. Sust. Energ. Rev.* **130** 109960.

[4] Aresti L, Christodoulides P, Michailides C and Onoufriou T 2023 Reviewing the energy, environment, and economy prospects of Ocean Thermal Energy Conversion (OTEC) systems. *Sustain. Energy Technol. Assess.* **60** 103459.

[5] Cao J Y, Zheng L, Zheng Z Y, Peng J Q, Hu M K, Wang Q L and Leung M K H 2023 Recent progress in organic Rankine cycle targeting utilisation of ultra-low-temperature heat towards carbon neutrality. *Appl. Therm. Eng.* **231** 120903.

[6] Liu W M, Xu X J, Chen F Y, Liu Y J, Li S Z, Liu L and Chen Y 2020 A review of research on the closed thermodynamic cycles of ocean thermal energy conversion. *Renew. Sust. Energ. Rev.* **119** 109581.

[7] Wang H X, Lei B and Wu Y T 2023 Control strategies of pumps in organic Rankine cycle under variable condensing conditions. *Appl. Therm. Eng.* **234** 121226.

[8] Imre A R and Ahmed A M 2023 Effect of the working fluids critical temperature on thermal performance for trilateral flash cycle and organic Rankine cycle. *Inter. J. Thermofluids* **20** 100417.

[9] Krempus D, Bahamonde S van der Stelt T P, Klink W, Colonna P and De Servi C M 2024 On mixtures as working fluids of air-cooled ORC bottoming power plants of gas turbines. *Appl. Therm. Eng.* **236** 121730.

[10] Rosard D D 1980 Generalized Parameters for selection of turbines and working fluids for OTEC power systems. *J Eng Gas Turbine Power*. **102** 215-222.

[11] Stoecker W F 1994 Comparison of ammonia with other refrigerants for district cooling plant chillers. *ASHRAE Trans*. **100** 1126-1135.

[12] Kim N J ,Shin S H and Chun W G 2006 A study on the thermodynamic cycle of OTEC system. *J. Korean Solar Energy* **26** 9-18.

[13] Sun F M, Ikegami Y, Jia B and Arima H 2012 Optimization design and exergy analysis of organic Rankine cycle in ocean thermal energy conversion. *Appl. Ocean Res*. **35** 38-46.

[14] Yang M and Yeh R 2014 Analysis of optimization in an OTEC plant using organic Rankine cycle. *Renew. Energ.*. **68** 25-34.

[15] Yoon J I, Son C H, Baek S M, Kim H J and Lee H S 2014 Efficiency comparison of subcritical OTEC power cycle using various working fluids. *Int. J. Heat Mass Transf.* **50**: 985-996.

[16] Elmaihy A, Rashad A, Elweteedy A and Nessim W 2023 Energy and exergy analyses for organic Rankine cycle driven by cooling water of passenger car engine using sixteen working fluids. *Energy Convers. Manag.-X.* **20** 100415.

[17] Ngangué M N, Lekané N N, Njock J P, Sosso O T and Stouffs P 2023 Working fluid selection for a high efficiency integrated power/cooling system combining an organic Rankine cycle and vapor compression-absorption cycles. *Energy* **277** 127709.

[18] Qyyum M A, Khan A, Ali S, Khurram M S, Mao N, Naquash A, Noon A A, He T B and Lee M 2022 Assessment of working fluids, thermal resources and cooling utilities for Organic Rankine Cycles: State-of-the-art comparison, challenges, commercial status, and future prospects. *Energy Convers. Manag.* **252** 115055

[19] Xu W C, Zhao R K, Deng S, Zhao L and Mao S S 2021 Is zeotropic working fluid a promising option for organic Rankine cycle: A quantitative evaluation based on literature data. *Renew. Sust. Energ. Rev.* **148** 111267.

[20] Li C Y, Pan L S and Wang Y Z 2020 Thermodynamic optimization of Rankine cycle using CO2-based binary zeotropic mixture for ocean thermal energy conversion. *Appl. Therm. Eng.* **178** 115617.

Pilot project on "green" hydrogen generation at the territory of Turkmenistan

A.Ya. Jumayev*

State Energy Institute of Turkmenistan, Turkmenistan

a.jumayev.tm@gmail.com

Abstract. This paper recommends the production of the "green" hydrogen at the territory of Turkmenistan. The electrical energy required for the production of "green" hydrogen is generated by a photovoltaic solar station. Current work presents a modelling project, which consists of electric power source – photovoltaic solar station (PVS) with a capacity of 100 MW, system of generation of hydrogen (electrolyzer with a capacity 50 MW) and the system of freshening – installation of the reverse osmosis with an output of 80 tons water a day.
Keywords: Photovoltaic solar station, electrolyser, hydrogen energy, a reverse osmosis unit.

1. Introduction

Accomplishments of the goals of Paris agreement on climate; reduce of the greenhouse vapors' emissions and the decrease of carbonic fuel consumption - serve as a stimulus of hydrogen energy development. Hydrogen used for accumulation, conservation and transportation of the energy and at the same time provides a great potential for the development of low carbonic economy that in turn reduces man (anthropgennic) impact on climate change.

Advantages of hydrogen are the followings:

- Non-emission of carbonic dioxide as an energy carrier use,
- High power intensity (duty) (three times higher in comparison with natural gas)
- Production potentiality from varios sources including water, hydrocarbons and organic materials.

It is expected that key role in the process of energy system's (energy transition) global transformation, first of all related with decarbonization and low carbonic development (progress), would play "green" hydrogen. In addition, "green" hydrogen considered not only as an energy carrier for the final consumption, but also as a mean for energy storage and accumulation as well as load equilibrium (balancing) of power grids (taking into account the instability use of electric energy and its generation from alternating (variable) renewable energy sources) [1].

Road map, on the development of the international cooperation in the field of the hydrogen energy for 2022-2023 years, was adopted by the decree of the President of Turkmenistan on January 28, 2022. Within the scope of this road map, it is planned to realize 18 projects and activities, particularly, to work out the national strategy for the development of the hydrogen energy in Turkmenistan [2].

2. Objectives of pilot project:

- Analyse current condition and perspectives of the hydrogen energy development;
- Examine the opportunities of the realization of the pilot project on generation of hydrogen at the territory of Turkmenistan;

• Choose an appropriate place for the project construction;
• Consider the use of the renewable sources to provide unit's consumers with electricity;
• Investigate existing methods of generation of hydrogen and choose the apt one to apply at the territory of Turkmenistan;
• Calculate electric energy capacity, produced by the chosen source and consumed elements in generation of hydrogen system ;
• Set a cost for pilot project realization and determine a term of its recoupment (payback).

Turkmenistan has a huge potential for the generation of hydrogen production. Pilot project considers the construction of the two photovoltaic solar station (PVS) s installed with a capacity of 100 MW each in settlements in Mary (Serhetabat) and Lebap (Kerki) districts. They can serve as an energy source for generation of "green" hydrogen. If electrolyzers used jointly with photovoltaic solar station for current goal, then there is no need for use of special equipment both for produced energy conversion (generators, inverter), and for synchronization with total energy system. Water, besides electric energy, serves as a raw material for generation of hydrogen with the means of eletrolyzer installation (device). Electrolyzer's work efficiency and reliability rate directly depends on decontamination (purifying) degree. There is an opportunity to use non-fresh water in the technical way of generation of hydrogen, for this, a water-desalinating plant is needed. This can lead to the appreciation of the generation of hydrogen. Realization of the activities mentioned above (water preparation to the hydrogen production) costs 1 US dollar per m^3 or approximately 0.01 US dollar per kg of hydrogen on average. The process of electrolyzer, at perfect conditions, demands 9 kg of water to produce 1 kg hydrogen. These data were used in the analysis part [3,4,5,11].

In the pilot project, the issue of connecting the PV station to the general power supply network is emphasized. This paper [6] introduced two PI-PSO controllers to stimulate a DVR for enhancing the on-grid hybrid system under misfire and fire-through faults. The PI-PSO controllers' goal is to force the system voltage at the POT to control the voltage between the DVR and the load and consequently enhance the performance of the system during misfire and fire-through faults. The proposed PI-PSO controller showed a better performance concerning the system voltage, RESs' powers, and the WECS speed, electromechanical torque, and power. The proposed controller for the DVR also succeeded at keeping the continuous connection of the PV/wind/fuel cell to the system without disconnection for longer fault clearing times. The implementation of the hybrid system with the DVR creates an advantage that can ameliorate the efficacy of the on-grid HPS. In the future, this work may be implemented in a real hybrid power system.

3. Materials and methods:

- According to data, derived from open sources, method of analysis of the electric energy source's amount; also the method of pilot project's realization cost and its recoupment were employed to achieve the set goals[7,8,9,10];
- This work carries out project simulation, which consists of electric power source – photovoltaic solar station (PVS) with a capacity of 100 MW, system of generation of hydrogen (electrolyzer with a capacity 50 MW,) and the system of freshening – installation of the reverse ocmosis with an output of 80 tons water a day;
- Analysis of various types of electrolyzers was undertaken.

4. Results and discussion

To assess the prospects of an interaction of the solar energy source and electrolyzer in the generation of hydrogen at the territory of Turkmenistan, photovoltaic solar stations installed with a capacity of 100MW each in settlements in Mary (Serhetabat) and Lebap (Kerki) districts were chosen as the main source of electric power. It is observed that only electrolyzer and water-desalination plant are involved in photoelectric solar plants. Technical data of electrolyzer with a capacity of 1 MW were used in an analysis (table1).

The calculations of electric power's quantity generated in photoelectric solar plants per year carried out as following:

$$E = E_{udel,\beta} \cdot S = \sum_{i=1}^{12} E_{i,\beta} \cdot \eta_p \cdot \eta_{inv} \cdot \eta_m \cdot S \quad (1)$$

where E – annual energy production of PVS ; $E_{udel,\beta}$ – specific production PVS with a slope β per year; S – PVS's area; $E_{i,\beta}$ - solar energy receipt to optimal oriented area with an angle $\beta = 36^o$ at an array latitude of PVS in i-th month. Losses of η_p in PSP is up to 25%, and efficiency η_{inv}direct current to alternating current is 98,8%, accepted efficiency η_m 19,2 % solar component cell.

Table 1. Technical characteristics (features) of electrolysis installation with power of 1 MW.

Technical features	**Value and measurement unit**
Nominal power	1 MW
Generation of hydrogen	300 Nm^3/h, 27kg/h
Control of hydrogen output	15-100%
Specific consumption of electric power	4,4 kW·h/Nm^3, 48,88 kW·h/kg
Hydrogen's outlet pressure	30-200 kgs/sm^2
Hydrogen's specific outlet	0,08988 kg/Nm^3
Lower heating power	119,96 MJ/kg (33,32 kW·h/kg or 3,00 kW·h/Nm^3)

During calculations, following presumptions were accepted:

- Working hours of electrolyzer with maximum mounted output in a day: January, February, December – 3 hours, march, april, may, September, October, November – 5 hours, june, july, august – 6 hours;
- Total amount of solar energy coming to the photovoltaic solar stations surface located in the settlement with an angularly slope $\beta=36^0$ of south orientation: Atamyrat (Kerki), Lebap province: 1919,328 kW·h/m^2 per year; Serhetabat (Kushki), Mary province: 1892,972 kW·h/m^2 per year.

Power of electrolyzer designed with a condition that PVS would cover demands in electric power. Therefore, generation of hydrogen would depend on output and mode of operation of PVS. Hence, it was decided to use electrolyzer with a powe or 50 MW.

Hydrogen's mass was calculated using the data, given in the table 1 (figure 2), derived or taken from electrolyzer with corresponding power of :

$$m = \sum_{i=1}^{12} m_i = \sum_{i=1}^{12} \eta_{el} \cdot t_i \cdot N_i \quad (2)$$

where $\boldsymbol{m}$ – mass of hydrogen produced in a year; m_i – mass of hydorgen, generated by electrolyzer in i-th month; η_{el} – work efficiency of electrolyzer; t_i – working hours of electrolyzer with maximum mounted output in a day; N_i – number of days in month.

Balance of generated and consumed energy of system components used in generation of hydrogen defined in (figure 1).

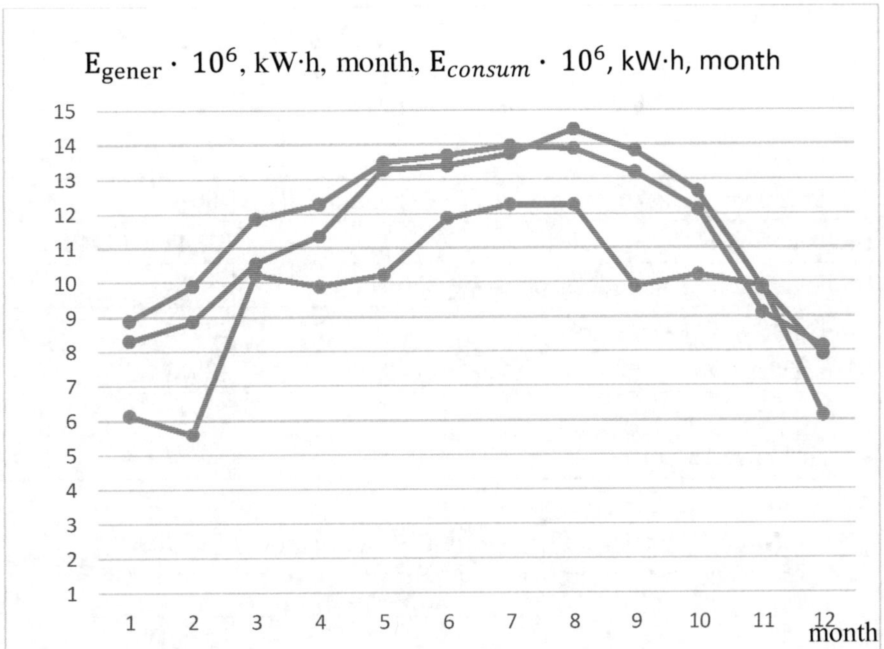

Figure 1. Generated electricity of PVS and consumed energy of electrolyzer (red and blue is PVS's generated energy; green – electrolyzer's consumed energy).

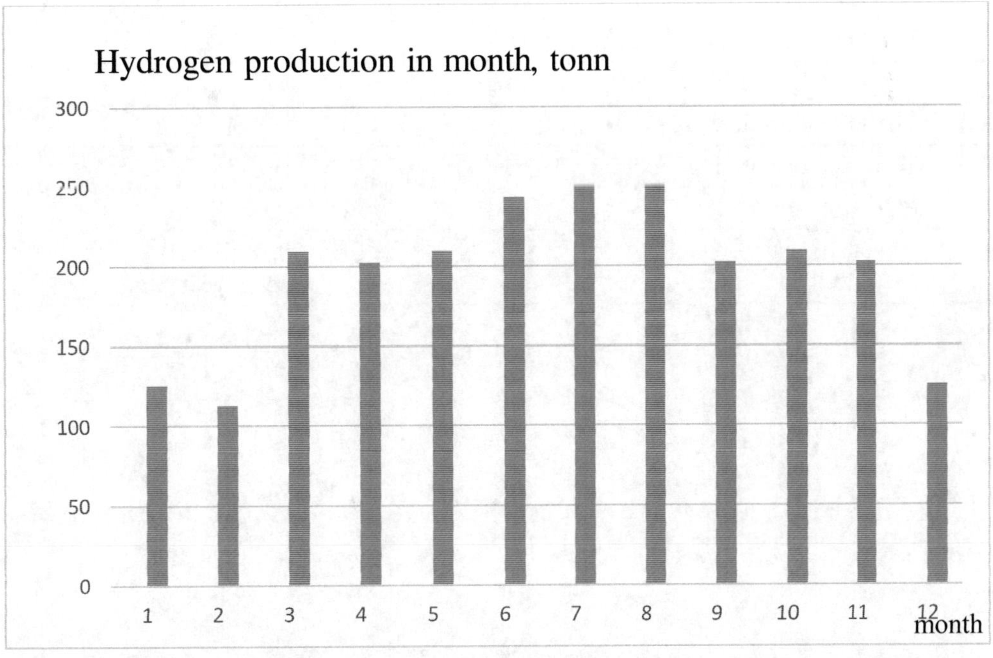

Figure 2. Hydrogen production in month.

Table 2 and table 3 show technical features of basic blocks station and PVS with a power of 100 MW. A scheme of basic blocks station shown in figure 3.

Journal of Physics: Conference Series **2723** (2024) 012010 doi:10.1088/1742-6596/2723/1/012010

Table 2. Technical features basic blocks.

1	PVS power	2745x2=5490 kW
2	Maximum power (P_{max}) photovoltaic (PV) module	380 W
3	Total number of PV modules	7224x2=14448
4	Total number of PV modules per string	28
5	Total number of strings	258x2=516
6	PV modules size	28578 m^2
7	Total number of inverters	2
8	Nominal inverter power	2841 kVA

Table 3. Technical features of PVS with a power of 100 MW.

1	PVS power	100 MW
2	Total number of basic blocks	18
3	Power of basic block	5490 kW
4	Power of PV module	380 W
5	Total number of PV modules	260064
6	Total number of PV modules per string	28
7	Total number of strings	9288
8	Total number of inverters	36
9	PVS size	514406 m^2
10	Annual energy production	$138{,}538 \cdot 10^6$ kWh/year

Journal of Physics: Conference Series **2723** (2024) 012010 doi:10.1088/1742-6596/2723/1/012010

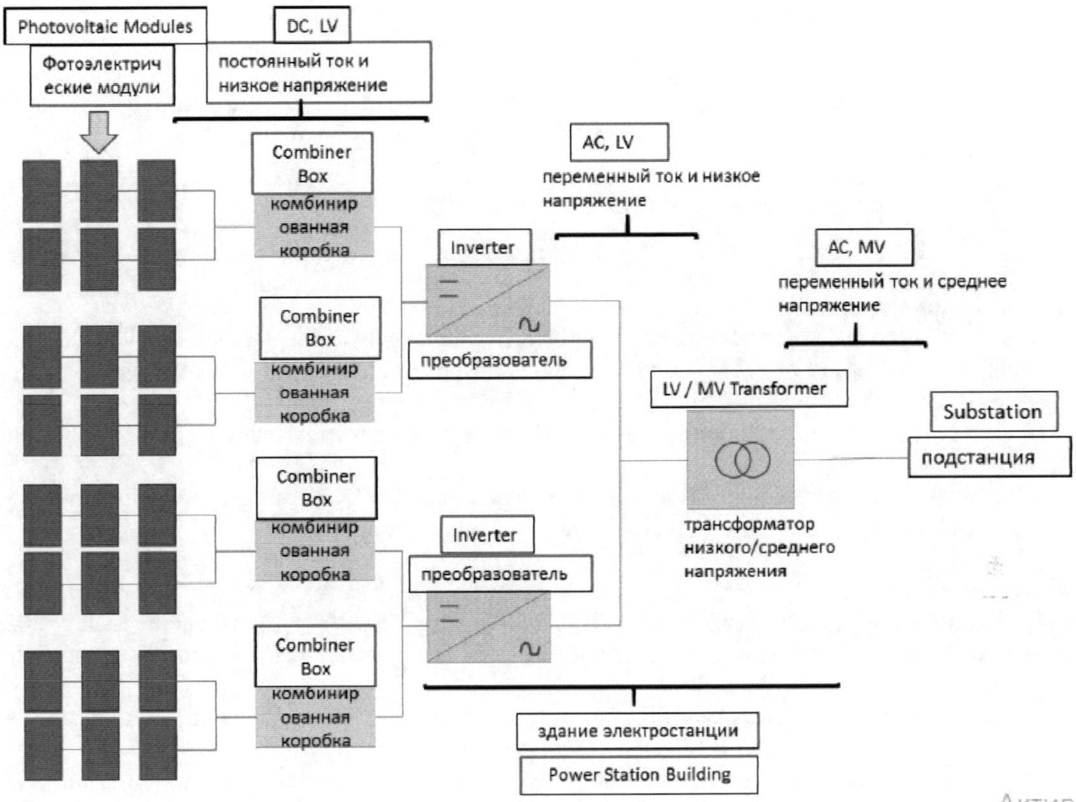

Figure 3. Scheme basic block.

Pilot project considers the construction of the two photovoltaic solar stations installed with a capacity of 100MW each in settlements in Mary (Serhetabat) and Lebap (Kerki) districts; and the derived results are shown in table 4. A realization of current pilot project enables to generate 2344,45 t of "green" hydrogen in a year.

Table 4. Basic results of pilot project.

Technical features and data	***Value and measurement unit***
Settlement, coordinates	Atamyrat (Kerki): northem latitude 37,8^0; eastern longitude 65,2^0
	Serhetabat (Kushki): northem latitude 35,2^0; eastern longitude 62,4^0
Total amount of solar energy coming to the solar panel's surface located in the settlement with an angularly slope β=36^0 of south orientation	Atamyrat (Kerki):1919,328 kW·h/m^2 year
	Serhetabat (Kushki):1892,972 kW·h/m^2 year
PVS's electric energy output (generation) per year	Atamyrat (Kerki):140,467·10^6 kW·h
	Serhetabat (Kushki):138,538 ·10^6 kW·h
Hydrogen generation	2344,45 t

Consumed energy in the generation process of hydrogen	$114{,}443 \cdot 10^6$ kW·h
Consumed water in the generation process of hydrogen	21104,550 t

5.Conclusions

1. Recommend to realize a pilot project on "green" hydrogen generation at the territory of Turkmenistan.
2. This work carries out project simulation, which consists of electric power source – photovoltaic solar station (PVS) with a power of 100 MW, system of hydrogen production (electrolyzer with a capacity 50 MW,) and the system of freshening – installation of the reverse ocmosis with an output of 80 tons water a day.
3. A realization of current pilot project enables to generate 2344,45 t of "green" hydrogen in a year. to create an empty row at the bottom of the table to contain them.

References

[1] Economic and Social Council. UN. 2022 ECE/Energy/2022/8. Distr.: General 13 July 2022
[2] Road map Turkmenistan, on the development of the international cooperation in the field of the hydrogen energy for 2022-2023 years. 2022
[3] Jumayev A Y 2022 The possibilities of using solar energy in the regions of Turkmenistan. Vestnik Gomel STU named after P.O. Suhoy. *Scientific-technical journal.* **3/4** (82,83),74-80.
[4] Jumayev A 2022 Distribution of solar energy potential on territory of Turkmenistan. *IOP Conf. Series: Earth and Environmental Sciences* 012157
[5] Jumayev A and Atayev E 2022 Pilot projects of industrial photovoltaic solar stations. *IOP Conf. Series: Earth and Environmental Sciences* 012142
[6] Elraouf M O A, Aljohani M, Mosaad M I and AbdulFattah T A 2022 Mitigating Misfire and Fire-through Faults in Hybrid Renewable Energy Systems Utilizing Dynamic Voltage Restorer. *Energies* **15**(16), 5769
[7] Bekirov E A, Asanov M M and Nusretova S S 2021 Development of energy complexes for the production of hydrogen in the territory of the Republic of Crimea. *Energy problems.* T **23**(2) 161-172.
[8] Filimonova A A, Chichirova A A, Chichirova N D, Filimonov A G and Pechenkin A V 2020 Prospects for the development of hydrogen power engineering in Tatarstan. *Power engineering research equipment, technology* **22**(6) 79-91
[9] Nazarow S and Jumaýew A 2006 The possibilities of using solar energy in welayats of Turkmenistan. *Turkmenistan science and technology* **6** 34-36.
[10] Mhitaryan N M , Kudya S O, Yatsenko LV, Shynkarenko L Y, Tkalenko M D and Budko V I 2013 Complex use of renewable energy sources. *International Scientific Journal for Alternative Energy and Ecology.* **17**(139) 14-22.
[11] Jumayev A Y 2022 Pathways to low carbon energy in Turkmenistan. *Materials XV International scientific-technology conf. Gomel STU named after P.O. Suhoy.* 321-324.

Comparison Study by using Pyrolysis of Kenna Sugarcane Bagasse and Sawdust in Sudan

Abuelnuor A.A. Abuelnuor [1,2]

[1] Department of Mechanical Engineering, Faculty of Engineering, Al-Baha University, Al Baha, Saudi Arabia.
[2]Sudan University of Sciences and Technology, Khartoum, Sudan.

abuelnuor99@gmail.com

Abstract: Today the request for save energy and addressing emissions has become a primary concern for every country aiming to achieve sustainable development while preserving its environment. The demand for energy is steadily rising, driven by rapid population growth and industrial development. Fossil fuels continue to play a major role in fulfilling these energy needs. , but their combustion is linked to rising environmental issues. There are many sources of energy use to solve the environmental problems and fossil fuel shorted such as biomass. This paper aims to explore the potential of biomass to provide significantly higher amounts of useful energy while reducing pollutant emissions compared to fossil fuels. In this research the two samples of sugar cane bagasse and sawdust is study by slow pyrolysis. Two samples are presented—one from Kenana bagasse and the other from sawdust. Bio-oil is a product of the slow pyrolysis process, and its calorific value, determined through laboratory analysis, is significantly high compared to previous studies. This indicates that bio-oil emits fewer pollutants than fossil fuels, making it suitable for use in transportation and various industrial sectors.
Keywords— pyrolysis process, sugarcane bagasse, sawdust, bio-oil, bioenergy production

1. Introduction

Energy is the foundation of life. There is several activities a moment that is unrelated to energy in life. The energy is using every moment of the day. In ancient times, humans relied on muscle power, later progressing to the use of fire and animal power. Subsequently, throughout history, humans have acquired the skill of harnessing energy and transforming it into a practical form for diverse application. Now a day there are primarily three sources of energy: fossil fuels, alternative energy such as nuclear and renewable. The conventional energy sources (fossil fuels) are coal, oil, and natural gas; the sources of renewable energies such as wind, solar, geothermal, hydropower and alternative fuel such as biomass biogas and biofuel and the alternative energy such as nuclear[1, 2]. Now a day the Fossil fuels were represent around 80 % of total sources of energy in the world[3]. The construction of fossil fuel has not been stopped until now. The world consuming of the energy from the conventional fuel is more than it produced these led to depleting the fossil fuel storage.[4]. The disadvantage of burning conventional fuels is the environmental impact and unavailability. Though, during the burning of coal, natural gas and oil, the products of this reaction emitted emissions to the atmosphere and are called greenhouse gases.

Furthermore, the exhausts gases form the combustion of the fossil fuel with oxidizer absorb energy from the sun and cause ozone. Global warming and various pollutants are released into the atmosphere, soil, and water. These pollutants emissions take an affected on the environment and humans. Air pollution contributes to respiratory diseases such as asthma. The air pollutants contribute the acid rain [3]. For these reasons, researchers in universities and academic centres of industries were focusing to products exploring cleaner sources for alternative fuels aims to reduce greenhouse gas emissions and broaden the available energy resources [4]. The one advantaged of renewable energy will not run out, dissimilar energy from fossil fuels [5]. There are many sources of currently, various renewable energy sources are actively utilized, such as hydropower, solar, WTE, biomass, hydropower, wind, thermal energy storage, and geothermal [6]. In recent times, biomass fuel has gained increasing attractiveness as a viable alternative to fossil fuels. This is driven by the growing demand for clean energy, the diminishing reserves of traditional fuels, and its role in reducing dependence on crude [5]. Biomass has the capability to be converted into biofuel over different physical methods and thermal biological. The focus of this paper is on the pyrolysis process applied to sugarcane bagasse and sawdust. Pyrolyzing woody biomass and other solid organic wastes is a promising thermochemical process [6, 7]. The sugar manufacturing sector, particularly agricultural industries, generates significant amounts of sugarcane bagasse. After crushing and the process of obtaining juice from sugarcane, the fibrous waste remnants of cane stalks, known as bagasse, are produced.

The hue of sugarcane bagasse typically ranges Shifting from a gray-yellow shade to a light green color, displaying a non-uniform and bulky particle size. The accumulation of sugarcane bagasse represents an underutilized and sustainable agricultural resource, comprising two distinct cellular components. The first component consists of "a thick, comparatively extended, fibrous component originating from the rind and fibro-vascular bundles distributed within the interior of the stalk. The second component is a pith segment obtained from the thin-walled cells of the ground tissue [1, 8, 9]. The content of wet sugarcane bagasse for each one ton of wet sugarcane approximately 10% (hundred kg)of sugar, 3.5% (thirty five kg) molasses, and 27% (two hundred seventy kg) wet bagasse is produced[8, 10-14]. Both the upper (higher) calorific value and lower calorific value of biomass, like bagasse, are determined through numerical formulas that consider the percentage of main components such as Brix, moisture, and ash. The moisture content significantly impacts both the upper heating value (UHV) and lower heating value (LHV) of damp biomass, such as bagasse. Pyrolysis involves the chemical breakdown of biomass, utilizing heat energy to bring about alterations in composition. It can occur with minimal or complete absence of oxidation. In other words, it does not allow gasification, indicating that the conversion of biomass samples takes place solely under the influence of thermal energy.

Furthermore, the pyrolysis is not solely a thermal conversion process for biomass, then include direct burning of biomass and gasification. . During pyrolysis, the extensive hydrocarbon chains containing hydrogen, carbon, and oxygen present in intricate macromolecules of biomass undergo fragmentation into smaller, simpler molecules. This transformative process yields condensable vapors (tars or oils), solid residue (char), and gas [15-21] . The characteristics of these products, influenced by the pyrolysis conditions, vary in their thermochemical and physical properties. The primary aim is to achieve a high calorific value and optimal combustion properties, particularly a high H/C ratio, for the original biomass. Fast pyrolysis, which has gained significant attention recently, holds the potential to generate superior products. Pyrolysis can be categorized into slow and flash pyrolysis based on the heating rate intensity applied [22-27].

The aim of this paper to make comparison between pyrolysis of sugarcane bagasse and sawdust in University of Science and Technology in Sudan.

2. METHODOLOGY

2.1. Pyrolysis Process

Figure 1 show the test rig was used in this experimental study. The main components in the experimental setup for this study include the pyrolysis reactor, cyclone separator, cooling mechanism, and interconnecting conduits. Within the pyrolysis reactor, both bagasse and sawdust samples are positioned for combustion. Volatile particles exit the reactor through a tube linked to its upper portion. Constructed from high thermal resistance stainless steel, the pyrolysis reactor boasts dimensions of internal diameter of 275 mm, thickness of 8 mm, and a height measuring 40 cm. A cyclone separator is employed to segregate particles from the gas, while the cooling system aims to purify the syngas by eliminating condensable gases. The system utilizes a double-pipe heat exchanger (1-1 counter flow) with an inner diameter of 50 mm, a thickness of 25 mm, and a height of 450 mm. The hot gases from the reactor flow through the smaller tube inside the cyclone separator, while cooling water circulates in the annular space between the two tubes, facilitating the condensation of gases and allowing the incondensable gases to pass through. The role of the connecting pipes is to facilitate the transfer of flowing gases across all components of the system, requiring resistance to high temperatures. Steel pipes, with diameters ranging from 1 to 2 inches, are employed for this purpose. The length of these pipes measures 150 mm from the reactor to the cyclone and 500 mm from the cyclone to the cooling system.

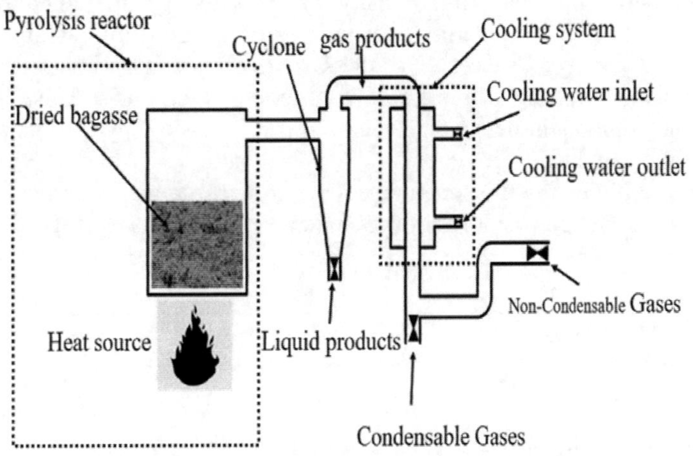

Figure 1. Pyrolysis process[1].

2.2 Procedure of Experimental

Samples preparation of sugarcane bagasse and sawdust. The bagasse was obtained from the Kenana Sugar Factory with very high content of moisture (50-52) %. The sawdust was sourced from a carpentry workshop at Sudan University of Science and Technology. Prior to the pyrolysis process, the sugarcane bagasse underwent a drying period of approximately ten days using direct sunlight and atmospheric air to decrease the moisture content to a range of 5-15%.Slow pyrolysis process was used in this experiment. The experiment took place in an external workshop, specifically set up locally and transported to the university campus. In a standard trial, briquettes of bagasse and sawdust samples were individually subjected to the same conditions. The conditions of the experiment was introduced the samples into the reactor at the temperature of room (40°C). The test commenced with the initiation of heating after ensuring that the air was sealed tightly and properly closed within the reactor. The subsequent steps of the test were then executed as follows:

State (1): A total of 0.75 kg of biomass of both samples was full into the reactor separately.
State (2): Following the loading of the samples, the reactor was completely sealed, and subsequently, the heating system was activated.
State (3): The Type-K thermocouple monitored the temperature increase over time, and after a few minutes, the vapor began to evaporate.
State (4): The initial experiment took a total of 50 minutes.
State (5): After a duration of 50 minutes, the external temperature of the reactor was measured by positioning a thermocouple on the upper part of the reactor cover.

State (6): A gas analyzer sensor was positioned at the gas outlet, and four results were documented at 5-minute intervals over four times, totaling 20 minutes. It's important to note that the syngas outlet was sealed after each reading.
State (7): Extracts from the generated tar and bio-oil were collected and sent to an external laboratory. The experiment was replicated for both bagasse and sawdust samples, conducted on two separate occasions.

3. Results and Discussions

In this experiment for pyrolysis of samples Kenna sugarcane bagasse and sawdust. For The Kenna sugarcane bagasse the results were as follow: A uniformly dark brown (semi-black) liquid, gas with a blue flame when burned and a black high-density tar. Additionally, there is dark brown bio-char with some substances that have not undergone decomposition, remaining within the reactor after the completion of the slow pyrolysis process. For the sample of sawdust a consistently dark brown (semi-black) liquid and a gas exhibiting a blue flame upon combustion and a dense black tar are produced. Table I illustrates the component percentages derived from the slow pyrolysis process for both Kenana sugarcane bagasse and sawdust.

Table 1. Table indicating the percentage of various components derived from the pyrolysis experiment for both Kenana bagasse and sawdust.

Samples	Char	Bio-oil	Tar	Gas + losses
Kenna bagasse	51%	8%	13%	27.9%
Sawdust	34.2%	7%	11%	47.8%

The mass distribution analysis of Kenana bagasse indicates that 51% is char, 8% is bio-oil, 13% is tar, and 27.9% accounts for gas loss. Meanwhile, for sawdust, the mass distribution is as follows: 56.2% char, 6.14% bio-oil, 8.8% tar, and 28.8% gas loss. A Kane gas analyzer was using for analyzed the exhaust, which measures different gas emissions such as carbon dioxide (CO_2), carbon monoxide (CO) and unburnt hydrocarbons (HC). The results of these emissions from pyrolysis of Kenana bagasse and sawdust samples are shown in Figures 2, 3 and 4. Figure 2 below illustrations and compare CO_2 emissions between the Kenana bagasse and sawdust samples with time. The X-axis expressions the time; Y-axis shows the emissions of CO_2 per volume. The results from this figure show that the volume of emitted CO_2 from pyrolysis of Kenana bagasse started with a high level of emission level, then it is reduced gradually. Though in the case of the sawdust, the emission of CO_2 started at a low level through a continuous rising with time. From these results, it is evident that the impact on CO_2 emissions is more pronounced in the case of sawdust. CO emissions from both Kenana bagasse and sawdust are depicted in Figure 3 below, with time represented on the X-axis and CO emissions per volume on the Y-axis. The CO emission level for Kenana bagasse initially started relatively high, experiencing a slight increase in the first few time intervals, followed by a gradual and continuous decline over time. In contrast, the CO emission level for sawdust started at a lower level and exhibited a continuous, somewhat sharp rise compared to the CO emission from Kenana bagasse.

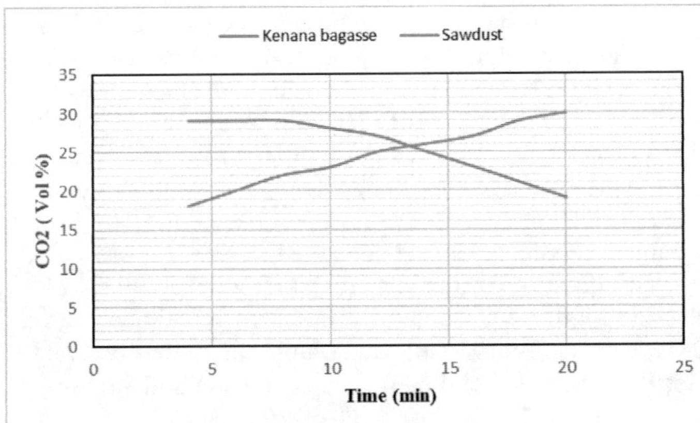

Figure 2. Concentrations of CO_2 obtained through the process of slow pyrolysis.

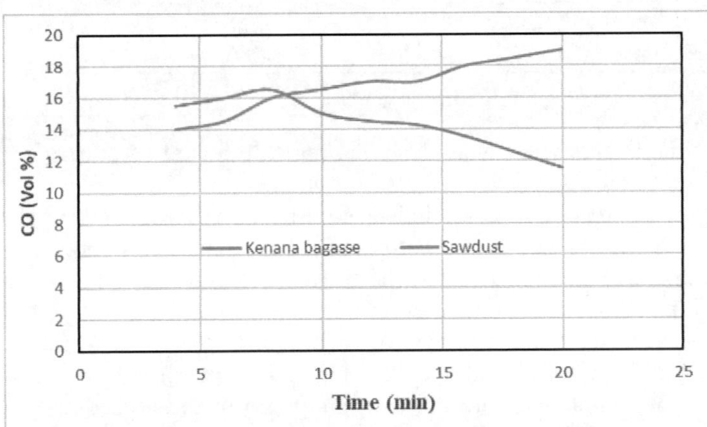

Figure 3. Concentrations of carbon monoxide (CO) achieved through the gradual pyrolysis process.

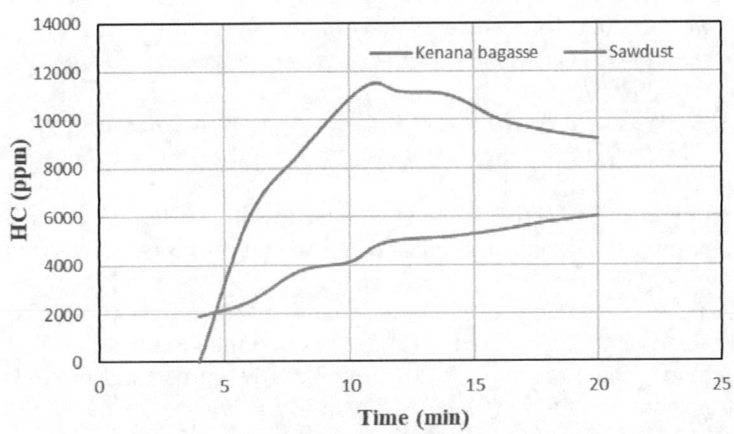

Figure 4. The rate of hydrocarbon (HC) production measured in parts per million (ppm) resulting from the slow pyrolysis process.

Figure 4 above, the graph illustrates the hydrocarbon (HC) emissions over time for both Kenana bagasse and sawdust samples. The X-axis represents time, while the Y-axis depicts hydrocarbon emissions in parts per million (ppm). For Kenana bagasse, HC emissions began slightly higher than those of sawdust, exhibiting a semi-sharp increase until reaching a peak around the midpoint of the experiment. Subsequently, there was a slight reduction until the end of the experiment. In contrast, sawdust HC emissions continued to rise throughout the test. HC emissions from both samples are crucial as they contribute significantly to the syngas produced during the pyrolysis process, characterized by a flame-producing ability. Igniting the produced syngas resulted in a clear blue flame. The gas analyzer used in the experiment provided the total of all hydrocarbons produced during pyrolysis.The observed variance in the percentage of syngas emissions between Kenana bagasse and sawdust can be attributed to their distinct compositions. The liquid oil produced from the slow pyrolysis of sugar cane bagasse and sawdust is further classified into bio-oil and tar. Thermochemical and physical property analysis, as presented in Table 2, reveals a cross calorific value higher than that reported in previous studies [29, 30]. This promising result is attributed to the low levels of water and humidity in the bio-oil.

Table 2. Heating value for Kenna bagasse and sawdust.

Samples	Cross calorific value (MJ/kg)
Kenna bagasse	41.67
Sawdust	40.6

4. CONCLUSION

The experimental setup for this study was designed and fabricated to adhere to the slow pyrolysis process. Once activated, the device yielded results from the slow pyrolysis of biomass, specifically sugar cane bagasse and sawdust. The outcomes included bio-oil, tar, bio-char, and syngas. Of these, the bio-char held particular significance due to the nature of the slow pyrolysis process, characterized by a voluminous production. Subsequent analysis of the generated syngas revealed that:

• The products emissions such as CO_2, CO, NO, and HC were measured.

•The examination of thermochemical and physical characteristics of bio-oil and tar indicated identical calorific values of 41.67 MJ/kg for bio-oil and 40.6 MJ/kg for tar. This rise in calorific value might be attributed to the low water content in bio-oil, as highlighted in the preceding point, along with low oxygen levels and reduced water content.

References

[1] Abuelnuor A A A, Mohamed A A, Omara AM, Rikabi I J A, Mohammed A S A, Almoez I A A and Moaz A A M H A 2020 Experimental study on pyrolysis of sugarcane bagasse for bioenergy production. *International Conference on Computer, Control, Electrical, and Electronics Engineering (ICCCEEE). IEEE.*

[2] Abuelnuor A A A , Wahid M A, Hosseini S E, Saat A, Saqr K M, Sait H H and Osman M 2014 Characteristics of biomass in flameless combustion: A review. *Renew. Sust. Energ. Rev*. **33**, 363-370

[3] Aravindan M and Kumar P J R i E 2023 Hydrogen towards sustainable transition: A review of production, economic, environmental impact and scaling factors. *Results in Engineering* **20**, 101456.

[4] MacCourt D C 2010 Renewable Energy Development in Indian Country: A Handbook for Tribes. National Renewable Energy Lab.(NREL), Golden, CO (United States).

[5] Juliet B I, Manovic Vand Longhurst P 2016 Biomass resources and biofuels potential for the production of transportation fuels in Nigeria. *Renew. Sust. Energ. Rev* **63**,172-192.

[6] Kolokolova O 2013 Biomass pyrolysis and optimisation for bio-bitumen. Theses / Dissertations

[7] Abuelnuor A A A, Gibreel M, Abuelnour M I, Younis O, Abuelgasim S and Mohmed E F 2023 Study of Energy and Exergy of Elgunied Sugar Company Plant in Sudan. *Sugar Tech* **25**, 1241–1250.

[8] Hassan H , Hameed B H and Lim J K 2020 Co-pyrolysis of sugarcane bagasse and waste high-density polyethylene: Synergistic effect and product distributions. *Energy* **191**, 116545.
[9] Rawat S S and Sharma A J M T P 2023 Sugarcane bagasse ash—The future composite material: A literature review. *Materials Today: Proceedings*. Available online 10 August 2023
[10] Hajiha H and Sain M 2015 The use of sugarcane bagasse fibres as reinforcements in composites, in Biofiber reinforcements in composite materials. B*iofiber Reinforcements in Composite Material.* 525-549.
[11] Garcìa-Pèrez M, Abdelkader C and Christian R 2002 Vacuum pyrolysis of sugarcane bagasse. *J Anal Appl Pyrolysis* **65**(2), 111-136.
[12] Varma A K and Mondal P J I C 2017 Products, Pyrolysis of sugarcane bagasse in semi batch reactor: Effects of process parameters on product yields and characterization of products. *Ind Crops Prod.* **95**, 704-717.
[13] Fioranelli A and Bizzo W A 2023 Generation of surplus electricity in sugarcane mills from sugarcane bagasse and straw: Challenges, failures and opportunities. *Renew. Sust. Energ. Rev.***186**. 113647.
[14] Osaki M R J I C 2022 An energy optimization model comparing the use of sugarcane bagasse for power or ethanol production. *Ind Crops Prod.* **187**, 115284.
[15] Adeniyi A G, Ighalo J O and Abdulsalam A 2019 Modeling of integrated processes for the recovery of the energetic content of sugar cane bagasse. *Biofuel Bioprod Biorefin (Biofpr)* **13**(4), 1057-1067.
[16] Ahmed N, Zeeshan M, Iqbal N, Farooq M Z and Shah S A 2018 Investigation on bio-oil yield and quality with scrap tire addition in sugarcane bagasse pyrolysis. *Ind Crops Prod.* **196**, 927-934.
[17] Tsai W T, Lee M K and Chang Y M 2006 Fast pyrolysis of rice straw, sugarcane bagasse and coconut shell in an induction-heating reactor. *J Anal Appl Pyrolysis* **76**(1-2):230-237.
[18] Montoyaa J I , Valdés C , Chejnea F, Gómez C A , Blanco A, Marrugo G, Osorio J, Castillo E, Aristóbulo J. and Acero J 2015 Bio-oil production from Colombian bagasse by fast pyrolysis in a fluidized bed: An experimental study. *J Anal Appl Pyrolysis* **112**, 379-387.
[19] Kar Y J B T 2011 Co-pyrolysis of walnut shell and tar sand in a fixed-bed reactor. *Bioresource Technol.* **102**(20), 9800-9805.
[20] Dewajani H , Zamrudy W, Irfin Z, Ningtyas D and Ridlo N M 2023 Utilization of Indonesian sugarcane bagasse into bio asphalt through pyrolysis process using zeolite-based catalyst. *Materials Today: Proceedings* **87**(2) , 383-389
[21] Wuri M A, Pertiwiningrum A, Budiarto R, Gozan M and Harto A W 2021 The Waste Recycling of Sugarcane Bagasse-Based Biochar for Biogas Purification. *IOP Conference Series: Earth and Environmental Science.. IOP Publishing.* **940** 012029
[22] Yogalakshmi K N , Poornima Devi T , Sivashanmugam P , Kavitha S , Yukesh K R ,Sunita V, AdishKumar S, Gopalakrishnan K and Rajesh B J 2022 Lignocellulosic biomass-based pyrolysis: A comprehensive review. *Chemosphere.* **286**,131824.
[23] Collard F Xand Joël Blin J 2014 A review on pyrolysis of biomass constituents: Mechanisms and composition of the products obtained from the conversion of cellulose, hemicelluloses and lignin. *Renew. Sust. Energ. Rev.* **38**, 594-608.
[24] Simkovic I , Francis B A and Reeves J B 1997 Pyrolysis-gas chromatography mass spectrometry analysis of starch-based ion-exchangers. *J Anal Appl Pyrolysis* **43**(2),145-155.
[25] Sohaib Q, Muhammad A and Younas M 2017 Fast pyrolysis of sugarcane bagasse: Effect of pyrolysis conditions on final product distribution and properties. E*nergy Sources, Part A: Recovery, Utilization, and Environmental Effects* **39**(2),184-190.
[26] Wang L , Yang Y, Ou Y, Zhong Q , Tang P , Zhang Y, Yi L ,Qian Li , Huang Z and Jiang T 2023 Synergistic recycling of biochar from sawdust pyrolysis and waste coke breeze to produce metallurgical quality biocoke with syngas as a by-product. *Fuel* **354**, 129365.
[27] Yanik J, Stahl R, Troeger N and Sinag A 2013 Pyrolysis of algal biomass. *J Anal Appl Pyrolysis* **103**,. 134-141.

Beyond Limits: A Brief Exploration of Fault Detection and Balancing in Lithium-ion Battery Technology

S S Madani and C Ziebert

Institute of Applied Materials-Applied Materials Physics, Karlsruhe Institute of Technology, 76344 Eggenstein-Leopoldshafen, Germany;

*Correspondence: seyed.madani@kit.edu

Abstract. The process of achieving balance among sequentially connected cells is crucial to prevent excessive charging or discharging, and it also improves the overall energy capacity. This article discusses various algorithms created for equalizing cell charge within a battery management system (BMS). Proper cell balancing is indispensable for upkeeping lithium-ion battery (LiB) packs. Within the BMS, identifying faults is of utmost importance. This encompasses detecting, isolating, and estimating faults. To prevent batteries from operating in unsafe ranges, it is vital to ensure the accurate functioning of current, voltage, and temperature sensors. Accurate fault diagnosis is pivotal for the optimal operation of battery management systems. In the context of electric vehicle battery management systems, precise measurement of current, voltage, and temperature is greatly relied upon to estimate the State of Charge (SOC) and overall battery health. Swiftly identifying early failures can mitigate safety hazards and minimize damage. Nevertheless, effectively pinpointing these initial failures using genuine operational data from electric vehicles remains a intricate task. This paper presents an analysis of different algorithms for detecting balancing-related faults, covering both methods based on models and those not reliant on models. The strengths and weaknesses of the evaluated algorithms, along with upcoming challenges in the realm of balancing and fault detection for LiBs, are also discussed in this document.

1. Introduction

LiBs find extensive application in new energy vehicles (NEV) due to their crucial role as substantial energy storage systems (ESS). For effectively managing and controlling such a vast number of cells, a precise and versatile battery management system (BMS) is absolutely essential. Within electric vehicles (EVs), LiB packs hold immense significance as a foundational element. These packs are composed of numerous cells interconnected using series and parallel setups to provide the necessary power and energy for the vehicle's operation.

In order to improve the effectiveness of techniques for diagnosing battery faults, Wang et al. [1] introduced a method for detecting faults in LiBs utilized within electric vehicles. This approach integrates multiple data-driven strategies to achieve a comprehensive analysis. Initially, deviations are eliminated, and initial fault assessments are conducted using a combination of t-distribution random neighborhood embedding (t-Sne) and wavelet transform denoising. Subsequently, significant vehicular attributes that have a notable impact on battery faults are identified through factor analysis. These critical attributes that lead to faults are then extracted using a combination of a two-way long and short-term memory network and a convolutional neural network. Finally, a self-learning Bayesian network is deployed to recognize and diagnose battery faults. Experimental results indicate that this method enhances the accuracy of fault diagnosis by approximately 12% when validated with various vehicle data. When compared to other approaches, this method not only achieves higher precision in fault diagnosis but also reduces the time required for fault identification. Furthermore, its performance surpasses that of graded fault systems, making it more closely aligned with real-world engineering applications [1].

Cai et al. [2] have introduced a multi-fault detection approach for battery packs connected in series. This technique incorporates elements of Attentional Mechanisms and Domain Adaptation Neural Networks, enabling the detection of various types of faults (specifically, voltage imbalance, internal short circuit, sensor anomalies, sensor drift voltage, and random fluctuation). By combining the advantages of both Attentional Mechanisms and Domain Adaptation Neural Networks, this innovative method, referred to as the Category-Reinforced Domain Adaptation Network, demonstrates effectiveness. It possesses the ability to diagnose multiple types of faults within battery packs. Through validation on a real-world platform encompassing three distinct operational scenarios, the approach showcases its capability. It enhances the model's ability to adapt across various battery pack contexts and notably contributes to the successful identification of multiple faults within battery packs [2].

Kosuru et al. [3] put forth a comprehensive framework for battery data analysis, with a specific focus on LiBs. This framework harnesses deep learning to detect and classify faulty battery sensor readings and transmission information. The study commences with the collection of sensor data, followed by preprocessing that involves z-score normalization. Subsequently, feature extraction is conducted using sparse principal component analysis (SPCA), while feature selection is facilitated through the enhanced marine predators algorithm (EMPA). A novel approach called the Incipient Bat-Optimized Deep Residual Network (IB-DRN) is introduced to enhance the safety and reliability of the BMS. This approach forms the foundation for identifying and categorizing erroneous battery data. The research employs MATLAB along with a combination of statistical analysis, machine learning techniques, and a deep learning toolbox. Experimental investigations are also conducted to demonstrate and evaluate the effectiveness of the proposed strategy. The results indicate that the suggested methodology surpasses conventional techniques, exhibiting superiority across various aspects.

Liu et al. [4] presented a solution in the form of a toggleable indicator designed to enhance the balancing of a battery pack connected in series. This is achieved by integrating a bypass equalizer with distinctive attributes including a compact topological structure, high efficiency, and intrinsic fault tolerance capabilities. The proposed toggleable indicator allows for the automatic selection of balance indicators, enabling a smooth transition between voltage and SOC based indicators. This inventive approach sets the stage for the development of a new balancing strategy that makes optimal use of the capabilities of the toggleable indicator. To validate the effectiveness of this approach, the proposed method undergoes both simulation and experimental testing using a LiB pack [4].

Fan et al. [5] introduced an MPC approach with rapid balancing to tackle inconsistency in LiB packs, optimizing energy transfer and minimizing SOC differences. Their MPC-based equalization algorithm aimed to efficiently reduce SOC variations, complemented by a quick-solving strategy to enhance computational efficiency. The method showcased superior performance in achieving swift equalization and minimizing energy wastage, with a key highlight being the avoidance of unnecessary battery cycling. This approach not only improved balancing accuracy but also demonstrated enhanced computational effectiveness compared to conventional techniques.

2. Balancing of LiBs

Figure 1 demonstrates an instance of battery cells being unbalanced. Several possible arrangements have been suggested to achieve balance in a battery bank, as described in various proposals [6]. All of these arrangements adhere to three fundamental methods, as shown in Figure 2. The first method involves discharging excess energy from battery cells. The next approach revolves around redistributing energy from cells with higher charge levels to those with lower charge levels. The last strategy involves controlling the charging or discharging current of each cell to ensure consistent SOC levels across all cells [7].

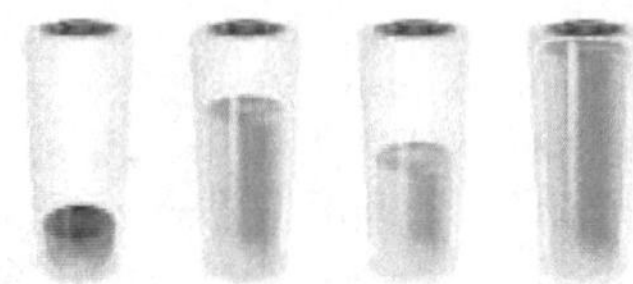

Figure 1. Unbalanced battery cells.

A highly reliable and efficient BMS is of great significance for applications powered by electrochemistry. Among the crucial features of a BMS, the aspect of cell balancing emerges as paramount. The purpose of cell balancing techniques is to ensure a fair distribution of energy among battery cells. The absence of proper cell balancing would lead to the wastage of capacity or energy within the battery array, a concern that becomes more pronounced in extended battery strings subject to frequent charge-discharge cycles. Qi et al. [7] categorizes several well-established cell balancing methods, organizing them based on their approach to managing excess energy within battery cells [7].

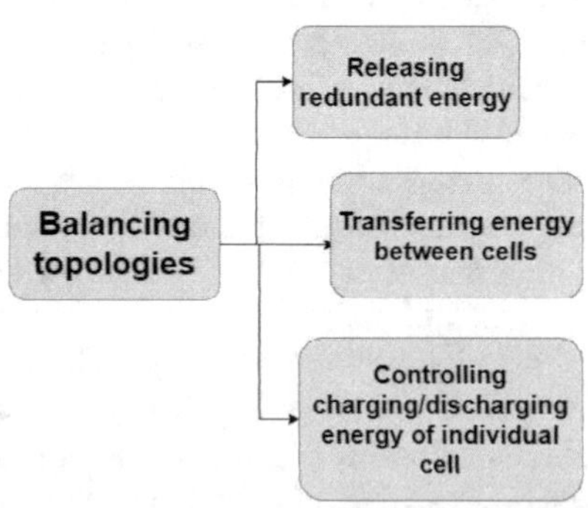

Figure 2. Cell Balancing Topologies Category [7].

Piao et al. [8] utilized a distance-based outlier detection method that relies on two main parameters—voltage and SOC—to calculate unusual values for individual cells. This enables the detection of unbalanced cells. By means of an online clustering approach, the algorithm differentiates between battery cells that are normal and those that are abnormal. To address the imbalance, bleeding circuits were applied to the cells identified as abnormal. The simulation results illustrate that the proposed balancing algorithm significantly improves the usable capacity of the battery pack, achieving a notable 9.5% increase compared to situations where balancing was not implemented [8].

Chowdhury and colleagues [9] introduced a novel strategy to achieve equilibrium in the State of Health (SOH) among interconnected LiB cells using DC/DC converters. SOH has become a significant gauge of battery health, alongside SOC, internal resistance, and conventional metrics. The adjustment of SOH balance is crucial for effectively utilizing battery cells that have varying capacity characteristics due to aging. They presented a simplified configuration to balance SOH in LiB cells linked via DC/DC converters, providing power to a shared load. Their uncomplicated approach was especially fitting for situations with limited battery maintenance access. The adoption of the proposed SOH balancing technique had the potential to elongate the overall operational lifespan of the battery storage unit and curtail replacement expenses.

Bonfiglio and colleagues [10] presented an active method for equalizing cells within stacks of LiBs. It is a widespread practice to balance the charge of individual cells in multi-cell LiB stacks to prevent harm and improve battery longevity. At present, the majority of battery stacks utilize a passive cell balancing approach that dissipates charge as heat through a resistor. In contrast, their approach utilizes a flyback converter to transfer charge between cells, leading to minimal energy wastage and a slight increase in cost compared to conventional passive systems.

McCurlie and his colleagues [11] introduced an advanced continuous-time fast model predictive control (MPC) method. This approach utilizes performance measures to establish an equilibrium in the SOC among battery pack components. Simulation outcomes illustrate that the MPC strategy achieves singular-point convergence of SOC, surpassing the performance of a traditional rule-based algorithm. This development enhances the efficiency of power electronics and extends the operational lifespan of battery cells by reducing frequent transitions between cell charging and discharging. Experimental results highlight a redistributive battery balancing system that rapidly accomplishes equilibrium by combining fast MPC with readily available microcontrollers found in today's market.

Zhang and co-authors [12] introduced a dynamic method for actively balancing the charging and discharging of LiB packs based on the average SOC. Two distinct active balancing strategies were developed to manage the varied charging and discharging conditions of the LiB pack. For charging, a balancing strategy was implemented that focuses on cells with SOC higher than the average SOC of the LiB pack, thereby enhancing the overall charging capacity. Conversely, during discharging or periods of rest, a different balancing strategy was employed, targeting cells with SOC lower than the average SOC of the LiB pack, thus improving the overall discharging capacity. Experimental findings confirm the effectiveness of this proposed active balancing approach. It showcases a reduction in energy discrepancies among battery cells and an improvement in both the charging and discharging capabilities of the LiB pack.

In contrast to prior studies, the maximum allowable current for cell equalization is adjusted to vary in accordance with changes in the external current of the battery pack. Ouyang and colleagues [13] aimed to prevent cell currents from surpassing their predetermined limits, as opposed to using a constant maximum value. By employing adaptive quasi-sliding mode observers to accurately SOC, they introduced a discrete-time quasi-sliding-mode-based strategy that integrates limited equalization currents. This strategy coordinates the collaborative operation of converters to efficiently achieve SOC equalization among cells. Through mathematical analysis and supported by experimental results, they demonstrated that the actual SOC discrepancies among cells can swiftly converge within an acceptable range centered around the origin.

To thoroughly investigate the influence of load cycles in conjunction with different BMS, Ziegler and co-authors [14] conducted a comprehensive assessment of individual cells. The findings of the study indicate that active balancing leads to a maximum improvement of 2% in usable capacity compared to passive balancing. Furthermore, active balancing alleviates inconsistencies among cells in terms of capacity and internal resistance. Notably, the study demonstrated that the additional stress caused by frequent energy redistribution in active balancing systems does not result in detrimental aging effects on cells when compared to passive balancing approaches.

Journal of Physics: Conference Series **2723** (2024) 012012 doi:10.1088/1742-6596/2723/1/012012

Räber and collaborators [15] provided a theoretical examination that details an approach for evaluating the benefits of cell-to-cell active charge balancing circuits compared to passive balancing alternatives, with a specific focus on energy preservation and capacity improvement. The assessment involves the utilization of variable parameters, including battery system configuration and distribution characteristics of cell capacity. Their study computed the inefficiencies linked to passive balancing within a battery system, while also approximating the overall energy conservation achievable through the use of cell-to-cell active balancing. The increase in capacity of a battery system utilizing active balancing, relative to a passive equivalent, was determined to fall in the range of 1.06 to 1.01.

Hua et al. [16] aimed to present a reliable and accurate method for assessing the SOC and SOH of battery packs that consist of multiple cells linked in a series configuration, utilizing passive balance control. In the initial stages of their research, the authors introduced the notions of individual cell-level and overall pack-level states. They provided clear explanations of how the conditions of individual battery cells are interconnected with the overall state of the battery pack. Subsequently, they formulated a multi time-scale approach to predict the SOC/SOH of the battery pack. Within this framework, the evaluation of SOH (which relates to gradual changes) is conducted over extended time periods, while SOC (which involves rapid changes) is estimated in real-time. To put their approach into practice, they adopted a non-linear predictive filter (NPF) as the estimation algorithm, which significantly enhanced the accuracy of SOC and SOH assessments. To validate their findings, the researchers performed experiments using a battery pack under specific driving cycles, confirming the effectiveness of the proposed methodology. The experimental results underscore the capability of the methodology to precisely gauge both the SOC and SOH of the battery pack.

3. Fault diagnostic of LiBs

The causes of battery deterioration at the anode and their consequences are depicted in Figure 3. Wu et al. [17] provided a comprehensive overview of the most recent research and progress in comprehending the mechanisms behind the aging of LiBs. By gaining insights into the origins and indications of battery malfunctions, this review encompasses the predominant factors that contribute to aging, along with the corresponding effects and results. Through the implementation of aging tests, the intricate relationship between performance and aging components, as well as their interconnected consequences, can be accurately gauged. The article concludes by summarizing recent advancements in diagnostic technologies for detecting battery malfunctions, accompanied by a fair assessment of their pros and cons. Ultimately, the paper proposes innovative strategies for diagnosing faults and outlines the enduring challenges that persist in this field [17].

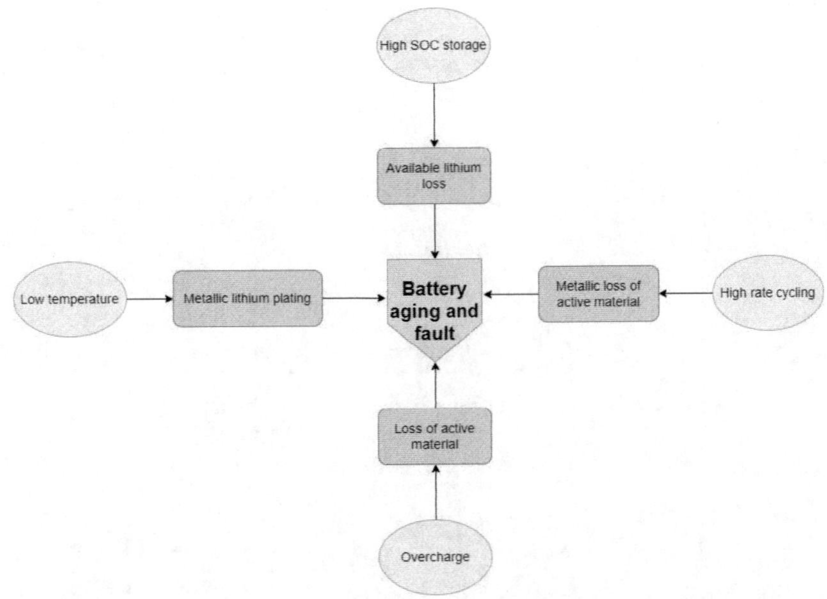

Figure 3. Causes for battery ageing at anode and their effects [17].

A basic diagram illustrating the algorithm used for identifying faults via state estimation is presented in Figure 4. Diverse categories of battery models are accessible, including those related to electrochemical, electrical, thermal aspects, and interdisciplinary variations such as electro-thermal combinations [18]. Figure 5 demonstrates transformations in a diagnostic system. The selection of a model for aiding in fault detection relies on the specific requirements of the LiB application. Model-based approaches are commonly preferred for fault detection due to their simplicity and cost-effectiveness. These approaches encompass techniques like state estimation, parameter estimation, parity equations, and structural analysis [20]. The categorization of diagnostic algorithms for identifying faults in LiBs is depicted in Figure 6. Diverse industries utilize a range of approaches for diagnosing faults. Yet, concerning LiB applications, the presence of both internal and interconnected faults often renders many conventional techniques from other domains inapplicable. Within the scope of LiBs, methods for diagnosing faults can be divided into two main groups: those that utilize models and those that don't rely on models [19]. This section will showcase recent progress in the realm of fault diagnosis, addressing concerns related to both internal and external issues associated with LiBs. Broadly, the process of making diagnostic decisions can be conceptualized as a sequence of conversions or mappings applied to process measurements. Figure 5 illustrates the distinct transformations that process data undergo throughout the diagnostic process.

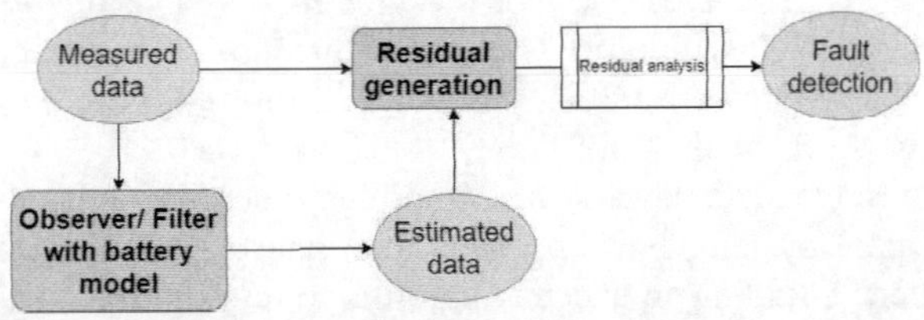

Figure 4. A simplified schematic of state estimation fault diagnosis [20].

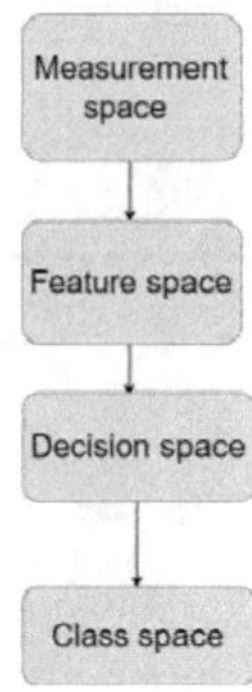

Figure 5. Transformations in a diagnostic system [19].

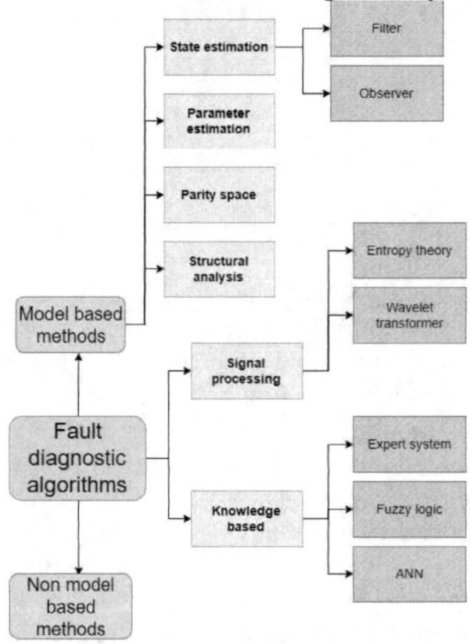

Figure 6. The classification of LiB fault diagnostic algorithms [20].

Xiong et al. [21] introduced a direct and efficient model-based strategy for detecting faults in sensors. Their method was specifically tailored to detect and isolate malfunctions in current or voltage sensors within an interconnected LiB pack. To differentiate between faults in sensors and faults in battery cells, a temperature sensor (assuming its proper operation) is utilized. This allows for distinguishing between a malfunction in a current or voltage sensor and a fault in a battery cell. Consequently, defective current or voltage sensors can be recognized by comparing the residual values of each cell with a predetermined threshold. The effectiveness of this proposed approach for diagnosing sensor faults is verified through both experimental tests and simulations, affirming its practical viability.

Zheng et al. [22] introduced an innovative technique to identify faults in voltage and current sensors within a LiB pack system. This approach integrates hybrid system modeling with the unscented particle filter, providing an efficient solution to this significant issue. By conducting experiments on a battery pack with a serial-parallel configuration, different fault scenarios concerning voltage and current sensors were evaluated. The results showcase that the method proposed in this study excels not only in precisely monitoring the system's state but also in accurately diagnosing faults within the sensors of the LiB system.

To extract insights related to faults from real-world electric vehicle usage data, Fan et al. [23] introduced an innovative approach involving the development of a generalized dimensionless indicator (GDI) along with a tolerance factor. The GDI is carefully crafted for the purpose of this study. By plotting these dimensionless indicators on a two-dimensional plane, the evolving patterns of voltage anomalies are visually displayed. To distinguish the changing anomaly patterns, the Local Outlier Factor (LOF) algorithm is utilized. This algorithm assists in recognizing the progression of anomalies, isolating malfunctioning batteries, and pinpointing the exact time of a fault occurrence. To mitigate the influence of cell variations on diagnostic accuracy, a differential technique is implemented. The effectiveness of the proposed method is validated through experiments conducted using real-world vehicle data. The results of the experiments emphasize the method's capability to effectively detect early-stage battery failures, showcasing both precision in fault identification and overall robustness.

Ma et al. [24] introduced an innovative dual-Kalman filter diagnostic method through a comprehensive analysis of the features of external soft-short circuit faults in series-connected LiB packs. The experiments were carried out for two main objectives: to understand the nature of the fault and to validate the effectiveness of the diagnostic algorithm. The suggested diagnostic technique exhibits remarkable capability in identifying external soft-short circuit faults, thereby highlighting its potential for precise fault detection and diagnosis.

Qiu et al. [25] utilized the local outlier factor (LOF) technique to conduct fault diagnosis on energy storage systems relying on LIB ESSs. The authors proposed two distinct algorithms for input generation within the LOF method: the multiple factors at a single time step input generation (MFST) algorithm and the single factor at multiple time steps input generation (SFMT) algorithm. To model various degrees of internal short circuit (ISC) severity, an ISC model was integrated into an electrical-thermal coupled model for an air-cooled LIB ESS. The performance of the LOF method was evaluated in detecting different levels of ISC severity, with assessments based on simulated data from the air-cooled LIB ESS and experimental data from a water-cooled LIB ESS.

Zhang et al. [26] presented a novel online multifault diagnosis approach that combines model-based and entropy methods to effectively detect and isolate various types of faults. These encompass faults in current, voltage, and temperature sensors, along with short-circuit faults and connection faults. To distinguish voltage sensor faults from battery short-circuits or connection faults, an interleaved voltage measurement topology is implemented. Leveraging a comprehensive battery model, structural analysis is employed to design diagnostic tests tailored to different fault categories. The extended Kalman filter is used for residual generation, followed by statistical inference for residual assessment, enabling the identification and isolation of sensor faults. Subsequently, sample entropy is employed to further differentiate between short-circuit faults and connection faults. The efficacy of the proposed diagnostic approach is validated through multiple fault tests involving various fault types and magnitudes. Importantly, the results highlight the method's resilience in managing noise and discrepancies in SOC and temperature data.

Lin et al. [27] introduced a comprehensive fault diagnostic strategy grounded in hybrid system theory, specifically designed to address common issues in LiB packs, including sensor and relay faults. This method utilizes automata, constructed on the principles of hybrid systems, to effectively incorporate the continuous and discrete dynamics inherent in battery packs. To manage computational complexity, a distributed diagnostic structure is employed, eliminating the necessity for a global battery pack model. The core of the approach lies in a dual extended Kalman filter algorithm, facilitating the estimation of crucial parameters like terminal voltages and SOC for individual battery cells within the pack. Residuals from current, terminal voltage, and SOC serve as foundational components for executing distinguishability analysis, which aids in identifying faults. The diagnostic process combines event observation with distinguishability analysis based on continuous dynamics. The efficacy of this diagnostic approach is verified through validation on a battery pack configuration comprising two series-connected batteries and two parallel-connected branches. Validation is conducted using the Federal Urban Driving Schedule driving cycle, affirming the robustness and practicality of the proposed method.

Xiong et al. [28] focused on online fault diagnosis specifically targeting External Short Circuits (ESC) in LiB packs. They conducted experimental investigations to gather and compare ESC characteristics between an NMC battery pack and a single cell. Utilizing the insights from experimental analysis, they formulated a two-step equivalent circuit model to describe the ESC process. An online model-based approach was then developed for diagnosing ESC faults within battery packs. The effectiveness of this approach was evaluated using experimental data. The results showcase its capability to precisely diagnose ESC faults within 3.5 seconds of occurrence, with a terminal voltage error of under 25 mV. Notably, the proposed method demonstrates robust generalization capabilities, successfully diagnosing ESC faults in battery packs with varying numbers of cells connected in series and in scenarios where current information is unavailable. In these cases, the terminal voltage error remains below 48 mV and 60 mV, respectively.

Jiang et al. [29] introduced a fault diagnosis technique for power LiBs, utilizing the isolated forest algorithm. The approach involves several key stages. Initially, the original voltage data undergo signal processing and decomposition, separating the data into static components, which exhibit strong correlations with aging inconsistencies, and dynamic components, which capture abnormal patterns. Subsequently, characteristic parameters are extracted from these static and dynamic components. These parameters are then fed into the isolated forest algorithm, which performs anomaly detection to identify cells with abnormalities. The proposed method was extensively tested using voltage data from four vehicles with defective batteries. The results of these tests confirm the method's ability to effectively detect both gradual and sudden failures. This validates its suitability for diagnosing power LiB faults and underscores its potential for real-time implementation in actual vehicles.

Shang et al. [30] introduced a real-time multi-fault diagnosis technique designed to detect early battery failures, utilizing a modified Sample Entropy approach. By analyzing the modified Sample Entropy of cell-voltage sequences within a moving window, the method effectively identifies and predicts various early battery faults, including short-circuit and open-circuit faults. Additionally, it can estimate the timing of fault occurrences. Experimental results, along with comparisons against traditional methods, validate the effectiveness of this proposed approach. It showcases robustness, reliability, and computational efficiency, all while bypassing the necessity for an exact model. In summary, this multi-fault diagnosis strategy demonstrates feasibility and potential in real-world electric vehicle applications, addressing the intricate nature of LiBs and offering early detection and prediction of faults.

Li et al. [31] introduced an inventive fault detection method that merges Empirical Mode Decomposition (EMD) with Sample Entropy (SE) to adeptly identify battery faults across diverse operating conditions. The proposed approach initiates by extracting relevant fault features using the EMD technique. This involves decomposing battery voltage signals and mitigating noise interference during voltage sampling. Through experimentation, the significance of fault features extracted by EMD is quantified. Subsequently, utilizing these extracted fault features, Sample Entropy values are computed. These values contribute to accurate fault detection and localization. Furthermore, a framework for evaluating the detected faults is devised to indicate the severity of battery faults. The efficacy of the method is then established using real-world data collected from electric vehicles, spanning scenarios involving routine and abrupt faults. This validation underscores the effectiveness of the proposed methodology in precisely detecting and diagnosing battery faults within varying operational contexts.

4. Challenges in Fault Detection and Balancing

Advancements in LiB fault detection include sophisticated BMS, integrating data analytics and machine learning for accurate fault prediction, enhanced sensor technologies, efficient thermal management systems, redundancy and diagnostics, and continuous condition monitoring. However, challenges persist in addressing complex failure modes, integrating advanced technologies seamlessly, balancing cost-effectiveness, ensuring data security, achieving scalability, and considering the full lifecycle of batteries. Overcoming these challenges requires collaborative efforts to develop standardized and cost-effective solutions, emphasizing ongoing research and development to ensure the safe and reliable use of lithium-ion battery technology across various applications.

Identifying issues and maintaining equilibrium are crucial aspects of effectively managing LiBs to ensure their safety, efficiency, and longevity. These batteries are widely used in various applications, from portable electronics to electric vehicles, and are susceptible to developing defects or inconsistencies over time that can compromise their performance and safety [32].

Here is a summary of fault detection and balancing in the context of LiBs:

Fault Detection: Detecting malfunctions in LiBs involves monitoring various parameters and characteristics to identify abnormal behavior [33]. Common malfunctions include excessive charging, extreme discharging, thermal escalation, capacity deterioration, and internal short circuits. Approaches to fault diagnosis include: Continuous Voltage and Current Monitoring: Persistent tracking of voltage and current during charging and discharging can reveal irregularities, such as abrupt spikes or plunges in voltage indicating problems [34].

Temperature Monitoring: Temperature sensors within the battery assembly can identify instances of overheating, a critical safety hazard [35].

Impedance Spectroscopy: Measuring the battery's impedance at different frequencies provides insights into changes in its internal structure and state [36].

Voltage Hysteresis: Monitoring voltage hysteresis during charge and discharge cycles can uncover capacity reductions or discrepancies [37].

SOC and SOH Estimation: Advanced algorithms can estimate the battery's SOC and SOH based on various parameters, helping discern deviations from normal behavior [38].

Balancing: Cell balancing involves equalizing the charge and discharge capacities of individual cells within a battery assembly [39]. Due to variations in production and operation, cells may have different capacities and voltage levels, leading to imbalances. Balancing is essential for maintaining consistent cell performance and extending the overall lifespan of the assembly. Various balancing strategies exist:

Passive Balancing: This strategy uses resistors or bleed circuits to disperse excessive charge from cells with higher voltages. It is simple but may be inefficient and generate heat [40].

Active Balancing: Active balancing incorporates additional circuitry to redistribute charge across cells, proving more effective and reducing heat production [41].

Cell Transfer Balancing: Physically transferring charge between cells is an alternative method, although less common due to its complexity [42].

Voltage Threshold Balancing: Balancing activates when a cell's voltage exceeds a specific threshold. This approach is efficient but requires monitoring circuitry for each cell [43].

Energy Transfer Balancing: Energy is transferred between cells using capacitors or inductors, proving more efficient and faster than some other methods [44].

To implement these techniques proficiently, battery management systems (BMS) play a crucial role [45]. A BMS oversees and regulates the battery's charging, discharging, and balancing processes. It collects data from individual cells and controls charging and discharging rates to prevent faults and imbalances.

Figure 7 illustrates the design of the Battery Management System (BMS) along with certain limitations. A significant challenge associated with the offline BMS is the inconsistency of state estimation algorithms [47]. Additionally, a notable drawback in the existing BMS is the absence of dependable real-time fault diagnosis algorithms [48]. Various battery faults encompass sensor malfunctions, cell connection issues, internal short circuits, external short circuits, overheating, and thermal runaway [49].

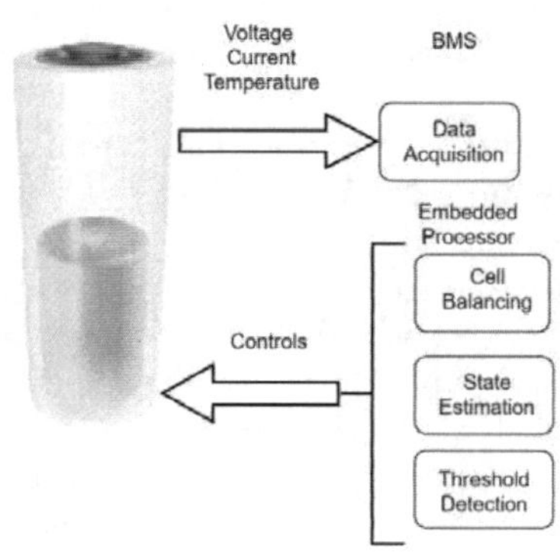

Figure 7. Battery management system design and functions [46].

5. Discussion

5.1 Introduction to LiBs and Challenges

LiBs have garnered increased attention as a dependable energy storage solution with extended operational lifespan and improved energy and power density. However, the natural aging of batteries leads to performance decline and potential system malfunctions, posing risks such as thermal runaway or explosions. Challenges arise from diverse production and operational conditions of connected cells, making it challenging for the BMS to ensure the safe operation of each cell.

5.2 Battery Balancing and Variability

Energy storage systems employing LiBs require cell balancing due to inherent manufacturing differences, varying self-discharge rates, disparities in internal resistances, and temperature variations among cells. Inconsistencies in LiB cell conditions can result in uneven residual capacity, risking cell damage due to overcharging or overdischarging.

5.3 Fault Diagnosis and Safety

Ensuring the safety and reliability of advanced battery management systems is a complex challenge, especially in diagnosing external soft-short circuit faults. Safety is paramount for LiBs,

Journal of Physics: Conference Series **2723** (2024) 012012 doi:10.1088/1742-6596/2723/1/012012

considering the potential for fire accidents under faulty conditions. Fault diagnosis involves detecting abusive conditions and identifying faulty batteries early to prevent escalation into thermal runaway. The LiB system faces various faults compromising performance and safety, and traditional fault-diagnosis methods struggle to identify early-stage issues without clear abnormality signs.

5.4 Challenges in Battery Management and Fault Detection
Voltage signals during battery operation are susceptible to noise interference. In electric vehicles, even minor malfunctions in power batteries can lead to accidents. Progressing safely in electric vehicles depends on swiftly identifying and accurately differentiating battery pack faults. Identifying fault characteristics within LiB packs, which vary in types and health statuses, presents challenges.

5.5 Sensor Data and BMS Robustness
Gathering and transmitting battery sensor data are crucial for BMS. Inaccuracies in battery data, stemming from sensor issues, communication problems, or cyber intrusions, can jeopardize BMS integrity and applications like electric vehicles. Hence, evaluating the robustness of battery sensor and communication data within BMS is critical. Sensor data underpin all BMS operations, and effective detection of sensor faults is pivotal for ensuring battery system sustainability and security in electric vehicles.

5.6 Battery Pack Balancing and Uniformity
Maintaining balance in battery packs is vital for their BMS. Insufficient balance limits overall battery pack performance to the capability of the weakest cell. Although Battery SOC is valuable for achieving balance, SOC estimation accuracy is not guaranteed, introducing uncertainties hindering balance optimization. Preserving LiB pack uniformity is essential for extending battery lifespan, maximizing capacity utilization, and ensuring safe electric vehicle operation.

Addressing these challenges requires collaborative efforts from researchers, industry stakeholders, and regulatory bodies to develop robust and standardized solutions for fault detection and balancing in lithium-ion batteries. Continuous research and development will play a crucial role in overcoming these challenges, ensuring the safe and reliable use of lithium ion battery technology across various applications.

6. Conclusion
The LiBs has gained increasing attention due to its potential as a dependable storage solution, with extended operational longevity and improved energy and power density. However, the battery's operation leads to inevitable aging, causing a decline in performance and potential system issues. These consequences not only bring inconvenience but also pose significant risks such as thermal runaway or even explosions. The Battery Management System (BMS) performs vital functions, including estimating battery state, equalizing cell voltages, managing thermal conditions, and diagnosing malfunctions. Challenges arise due to the diverse production and operational conditions of series and parallel-connected cells, making it difficult for the BMS to ensure the safe operation of each individual cell.

Energy storage systems utilizing LiBs require cell balancing due to inherent manufacturing variations, differing self-discharge rates, disparities in internal resistances, and temperature differences among individual cells. Variations in the conditions and parameters of LiB cells can lead to inconsistencies in remaining capacity among cells, potentially resulting in cell damage due to overcharging or overdischarging. Active charge balancing has gained recognition as an effective approach to create more energy-efficient and environmentally-friendly setups for LiBs.

Dealing with the variability among cells within battery packs for electric vehicles (EVs), accurately estimating the SOC and SOH of battery systems poses a significant challenge, especially within the computational limits of battery management systems. The reliable operation of electric vehicles heavily relies on the proper function of LiB packs. Early detection of potential failures in LiB packs is crucial to ensure the safe and reliable operation of these vehicles.

Ensuring the safety and reliability of advanced battery management systems is a complex challenge, especially in diagnosing external soft-short circuit faults. This specific issue remains intricate and unresolved. The safety of LIBs is of utmost importance, particularly due to the potential for fire accidents when they are faulty or operate under abnormal conditions. Ensuring the safety of LIBs is a critical requirement for their widespread use.

One method to achieve this is through fault diagnosis, involving the detection of abusive conditions and identification of faulty batteries at an early stage to prevent them from progressing to thermal runaway. The LiB system is vulnerable to various faults that can compromise its performance and safety. Detecting these early faults is difficult, and false alarms can occur due to similarities in fault characteristics. A dependable fault diagnostic method is essential to ensure battery system performance and safety.

Ensuring battery safety is crucial when using LiBs in all-climate electric vehicles. Issues like short circuits, overcharging, and overheating are prevalent. Incidents of electric vehicle safety stemming from LiB failures have become more frequent in recent years. Notably, voltage data from a malfunctioning battery often show abnormal patterns before a safety incident. These abnormal voltage changes are more pronounced in progressive failures, while sudden failures may involve subtler voltage shifts. Conventional fault-diagnosis methods struggle to identify battery faults early on when no obvious abnormalities are present. This challenge arises due to the complexity, nonlinearity, and time-varying nature of LiBs, compounded by inherent cell inconsistencies.

Fault detection is a critical aspect of LiB operation in electric vehicles. Voltage signals during battery system operation are susceptible to noise interference. Power batteries are central components in electric vehicles, and even minor malfunctions can lead to accidents. Therefore, ensuring accurate diagnosis of battery issues holds great importance.

Ensuring the intelligent advancement and safe operation of electric vehicles largely depends on swiftly identifying and accurately distinguishing various battery pack faults. However, pinpointing fault characteristics within LiB packs, which vary in battery types and health statuses, proves challenging. In this context, domain-adapted neural networks show promise.

Collecting and transmitting battery sensor data is of paramount importance in Battery Management Systems (BMS). Inaccuracies in battery data due to sensor malfunctions, communication errors, or even cyber intrusions can pose significant threats to both BMS integrity and applications dependent on BMS, such as electric vehicles. Thus, evaluating the resilience of battery sensor and communication data within BMS is a critical endeavor.

Sensor data underlie all BMS operations. Effectively detecting sensor faults is pivotal for ensuring the sustainability and security of battery systems in electric vehicles. Maintaining balance within a battery pack is crucial for its Battery Management System (BMS). When not adequately balanced, the overall performance of the battery pack is limited by its weakest cell. While Battery SOC is a useful indicator for achieving balance, the accuracy of SOC estimation is not always assured, introducing uncertainties that can hinder balance optimization. Maintaining uniformity in LiB packs is highly important to prolong battery lifespan, optimize capacity utilization, and ensure the safe operation of electric vehicles. In conclusion, fault detection and balancing are essential for ensuring the secure and optimal performance of LiBs. Regular monitoring, appropriate use of balancing techniques, and the application of advanced battery management systems all contribute to extending battery life and reducing safety hazards.

In conclusion, the widespread adoption of LiBs in energy storage and electric vehicles comes with challenges, particularly in ensuring safety and optimal performance. The Battery Management System (BMS) plays a crucial role in addressing issues such as cell balancing, fault detection, and accurate estimation of battery state. Challenges include the diverse conditions of connected cells, manufacturing variations, and the complexity of fault diagnosis. Active charge balancing and advanced fault diagnostic methods, including domain-adapted neural networks, show promise in enhancing energy efficiency and safety. Additionally, the resilience of battery sensor and communication data within BMS is vital. Overall, continuous advancements in monitoring, balancing techniques, and BMS applications are essential for extending battery life and reducing safety hazards in the dynamic landscape of LiB technology, particularly in electric vehicles.

Funding: This research was funded by the Helmholtz Association, grant number FE.5341.0118.0012, in the program Materials and Technologies for the Energy Transition (MTET). We want to express our gratitude for the funding.

Acknowledgments
This work contributes to the research performed at CELEST (Center of Electrochemical Energy Storage Ulm-Karlsruhe).

Reference

[1] Wang Z, Luo W, Xu S, Yan Y, Huang L, Wang J, Hao W, Yang Z, 2023. Electric vehicle lithium-ion battery fault diagnosis based on multi-method fusion of big data. *Sustainability*, **15**(2), 1120.

[2] Cai L, Wang H, Dong Z, He Z, Gao M, Song Y, 2023. A multi-fault diagnostic method based on category-reinforced domain adaptation network for series-connected battery packs. *J. Energy Storage*, **60**, 106690.

[3] Kosuru VSR, Venkitaraman AK, 2023. A smart battery management system for electric vehicles using deep learning-based sensor fault detection. *World Electr. Veh. J.* **14**(4), 101.

[4] Liu Y, Meng J, Yang F, Peng Q, Peng J, Liu T, 2023. A switchable indicator for active balance of the lithium-ion battery pack using a bypass equalizer. *J. Energy Storage*, **68**, 107696.

[5] Fan TE, Liu SM, Yang H, Li PH, Qu B, 2023. A fast active balancing strategy based on model predictive control for lithium-ion battery packs. *Energy*, 128028.

[6] Cao J, Schofield N, Emadi A, 2008. Battery balancing methods: A comprehensive review. In Vehicle Power and Propulsion Conference, 2008. VPPC '08. IEEE, pp. 1–6.

[7] Qi J, Lu DDC, 2014. Review of battery cell balancing techniques. In 2014 Australasian Universities Power Engineering Conference (AUPEC), pp. 1–6. IEEE.

[8] Piao C, Wang Z, Cao J, Zhang W, Lu S, 2015. Lithium-ion battery cell-balancing algorithm for battery management system based on real-time outlier detection. ***Math. Probl. Eng.*** (2)1-12.

[9] Chowdhury S, Shaheed MNB, Sozer Y, 2019. An integrated state of health (SOH) balancing method for lithium-ion battery cells. In 2019 IEEE Energy Conversion Congress and Exposition (ECCE), pp. 5759–5763. IEEE.

[10] Bonfiglio C, Roessler W, 2009. A cost-optimized battery management system with active cell balancing for lithium-ion battery stacks. In 2009 IEEE Vehicle Power and Propulsion Conference, pp. 304–309. IEEE.

[11] McCurlie L, Preindl M, Emadi A, 2016. Fast model predictive control for redistributive lithium-ion battery balancing. *IEEE Trans. Ind. Electron.*, **64**(2), 1350–1357.

[12] Zhang Z, Zhang L, Hu L, Huang C, 2020. Active cell balancing of lithium-ion battery pack based on average state of charge. *Int. J. Energy Res.*, **44**(4), 2535–2548.

[13] Ouyang Q, Chen J, Zheng J, Hong Y, 2017. SOC estimation-based quasi-sliding mode control for cell balancing in lithium-ion battery packs. *IEEE Trans. Ind. Electron.*, **65**(4), 3427–3436.

[14] Ziegler A, Oeser D, Arndt B, Ackva A, 2018. Comparison of active and passive balancing by a long term test including a post-mortem analysis of all single battery cells. In 2018 International IEEE Conference and Workshop in Óbuda on Electrical and Power Engineering (CANDO-EPE), pp. 000015–000020. IEEE.

[15] Räber M, Abdeslam DO, Heinzelmann A, Ramirez A, 2017. Performance estimation of a cell-to-cell-type active balancing circuit for lithium-ion battery systems. In 2017 IEEE 26th International Symposium on Industrial Electronics (ISIE), pp. 1005–1010. IEEE.

[16] Hua Y, Cordoba-Arenas A, Warner N, Rizzoni G, 2015. A multi time-scale state-of-charge and state-of-health estimation framework using nonlinear predictive filter for lithium-ion battery pack with passive balance control. *J. Power Sources*, **280**, 293–312.

[17] Wu C, Zhu C, Ge Y, Zhao Y, 2015. A review on fault mechanism and diagnosis approach for Li-ion batteries. *J. Nanomater.*, 2015, 8–8.

[18] Tomasov M, Kajanova M, Bracinik P, Motyka D, 2019. Overview of Battery Models for Sustainable Power and Transport Applications. *Transp. Res. Procedia*, **40**, 548–555.

[19] Venkatasubramanian V, Rengaswamy R, Yin K, Kavuri SN, 2003. A review of process fault detection and diagnosis. *Comput. Chem. Eng.,* **27**, 293–311.

[20] Tran MK, Fowler M, 2020. A review of lithium-ion battery fault diagnostic algorithms: Current progress and future challenges. *Algorithms,* **13**(3), 62.

[21] Xiong R, Yu Q, Shen W, Lin C, Sun F, 2019. A sensor fault diagnosis method for a lithium-ion battery pack in electric vehicles. *IEEE Trans. Power Electron.*, **34**(10), 9709–9718.

[22] Zheng C, Chen Z, Huang D, 2020. Fault diagnosis of voltage sensor and current sensor for lithium-ion battery pack using hybrid system modeling and unscented particle filter. *Energy*, **191**, 116504.

[23] Fan Z, Zi-xuan X, Wu Ming-hu, 2022. Fault diagnosis method for lithium-ion batteries in electric vehicles using generalized dimensionless indicator and local outlier factor. *J. Energy Storage*, **52**, 104963.

[24] Ma M, Duan Q, Li X, Liu J, Zhao C, Sun J, Wang Q, 2021. Fault diagnosis of external soft-short circuit for series connected lithium-ion battery pack based on modified dual extended Kalman filter. *J. Energy Storage*, **41**, 102902.

[25] Qiu Y, Dong T, Lin D, Zhao B, Cao W, Jiang F, 2022. Fault diagnosis for lithium-ion battery energy storage systems based on local outlier factor. *J. Energy Storage*, **55**, 105470.

[26] Zhang K, Hu X, Liu Y, Lin X, Liu W, 2021. Multi-fault detection and isolation for lithium-ion battery systems. *IEEE Trans. Power Electron.*, **37**(1), 971–989.

[27] Lin T, Chen Z, Zheng C, Huang D, Zhou S, 2020. Fault diagnosis of lithium-ion battery pack based on hybrid system and dual extended Kalman filter algorithm. *IEEE Trans. Transport. Electrification*, **7**(1), 26–36.

[28] Xiong R, Yang R, Chen Z, Shen W, Sun F, 2019. Online fault diagnosis of external short circuit for lithium-ion battery pack. *IEEE Trans. Ind. Electron.*, **67**(2), 1081–1091.

[29] Jiang J, Li T, Chang C, Yang C, Liao L, 2022. Fault diagnosis method for lithium-ion batteries in electric vehicles based on isolated forest algorithm. *J. Energy Storage*, **50**, 104177.

[30] Shang Y, Lu G, Kang Y, Zhou Z, Duan B, Zhang C, 2020. A multi-fault diagnosis method based on modified Sample Entropy for lithium-ion battery strings. *J. Power Sources* **446**, 227275.

[31] Li X, Dai K, Wang Z, Han W, 2020. Lithium-ion batteries fault diagnostic for electric vehicles using sample entropy analysis method. *J. Energy Storage* **27**, 101121.

[32] Tarascon JM, Armand M, 2001. Issues and challenges facing rechargeable lithium batteries. *Nature*, **414**(6861), 359–367.

[33] Zhang SS, 2011. A review on the separators of liquid electrolyte Li-ion batteries. *J. Power*

Sources **196**(1), 296–311.

[34] Uddin KMA, Davies PA, 2014. State of health estimation of Li-ion batteries in electric vehicles by electrochemical impedance spectroscopy. *J. Power Sources*, **272**, 1004–1013.

[35] Plett GL, 2015. Extended Kalman filtering for battery management systems of LiPB-based HEV battery packs: Part 3. State and parameter estimation. J. Power Sources, **273**, 1066–1075.

[36] Park J, Cho J, Kim Y, 2015. Thermal modeling and management of lithium-ion batteries. *J. Electrochem. Soc.***162**(14), A2538–A2550.

[37] Cao J, Liao Y, Lei Y, 2016. Review on state of charge estimation techniques for batteries employed in electric vehicles. *Renew. Sust. Energy Rev.*, **56**, 572–584.

[38] Zhu W, Onar OC, 2016. A review of state of charge indication methods for Li-ion batteries in electric and hybrid electric vehicles. *World Electr. Veh. J.* **8**(2), 469–479.

[39] Hu J, Li Z, Yang W, Pecht M, 2018. Prognostics and health management of lithium-ion batteries in electric vehicles: A review. *Energy Storage Mater.*, **10**, 246–257.

[40] Plett GL, 2018. Battery Management Systems Volume I: Battery Modeling. Artech House.

[41] Ouyang M, Lu L, Li J, 2019. A review on life cycle management of lithium-ion batteries. *J. Power Sources,* **435**, 226767.

[42] Wang C, Sun F, Tsui KL, 2014. A novel energy equalization method for series-connected battery string. *Appl. Energy*, **113**, 1640–1651.

[43] Wang Y, Wang C, Tsui KL, 2017. A comprehensive review of battery equalization methods for lithium-ion batteries. *Energy Storage Mater.*, **7**, 105–137.

[44] Wang C, Sun F, Tsui KL, Chan CC, 2013. Review of energy equalizers in rechargeable batteries. *IEEE Trans. Ind. Electron.*, **61**(3), 1154–1167.

[45] Liu H, Peng H, Li B, 2017. Advances in balancing techniques for lithium-ion batteries: State-of-the-art and future perspectives. *J. Power Sources*, **355**, 91–105.

[46] Tran MK, Panchal S, Khang TD, Panchal K, Fraser R, Fowler M, 2022. Concept review of a cloud-based smart battery management system for lithium-ion batteries: Feasibility, logistics, and functionality. *Batteries* **8**(2), 19.

[47] Shen M, Gao Q, 2019. A review on battery management system from the modeling efforts to its multiapplication and integration. *Int. J. Energy Res.* **43**, 5042–5075.

[48] Lu L, Han X, Li J, Hua J, Ouyang M, 2013. A review on the key issues for lithium-ion battery management in electric vehicles. *J. Power Sources* **226**, 272–288.

[49] Wu C, Zhu C, Ge Y, Zhao Y, 2015. A Review on Fault Mechanism and Diagnosis Approach for Li-Ion Batteries. *J. Nanomater.* 631263.

Study on synthesis and properties of polyacrylate wax inhibitor

Xiang Gao [1], Yuandan Zhang [1], Zhe Li [2], Xinmin Wang [1] and Yingna Du [3*]

[1]Changqing Oilfield No. 9 Oil Production Plant, China National Petroleum Corporation, Xi'an, Shaanxi, 710018, China
[2]Changqing Engineering Design Co., Ltd, China National Petroleum Corporation, Xi'an, Shaanxi, 710018, China
[3]Engineering Research Center of Oil and Gas Field Chemistry, Universities of Shaanxi Provence, Xi'an Shiyou University, Xi'an, 710065, China

*1294138205@qq.com

Abstract. Synthesis and properties of polyacrylate ester paraffin inhibitior based on polyacrylate ester start from dodecanol, tetradecanol, hexadecanol, octadecanol respectively were introduced in this paper. Hexadecyl polyacrylic showed the most potent paraffin inhibition, and then the optimum reaction condition of it was ascertained. The percentage inhibition of hexadecyl polyacrylic, synthesized under the optimized conditions, reached to 79.3%. Then crosslinked polyacrylate ester paraffin inhibitiors were synthesized and screened by static wax precipitation experiment, and find the percentage inhibition of crosslinked hexadecyl polyacrylic reached to 88.4%.
Keywords: Paraffin inhibition, Synthesis, Polyacrylate ester, Static wax precipitation

1. Introduction

The wax exists in the dissolved state in the formation reservoir. During the mining process, the waxy crude oil rises along the tubing. With the continuous decrease of pressure and temperature and the continuous escape of light components, the wax in the crude oil begins to crystallize and precipitate, and continues to deposit. When the crude oil temperature is lower than the critical cloud point temperature, the wax crystal molecules will diffuse to the solid surface, and begin to form a solid three-dimensional network structure with this as the center. The wax crystals are thin or needle-like adsorbed on the wall of the tube to form wax deposition. Wax deposition has a great influence on the production, gathering and storage of crude oil, which can lead to the decline of oil well production and even lead to shutdown [1]. Wax deposition is mainly divided into three processes: wax crystal formation, wax crystal growth and precipitation, and wax deposition. Controlling wax deposition in any step can prevent wax deposition in oil wells.

At present, there are three types of paraffin inhibitors used in oil fields: polycyclic aromatic hydrocarbon type, surface active agent type and polymer type. The non-polar part of the polymer type paraffin inhibitor crystallizes with wax, while the polar link distorts the crystal form of the wax crystal, which is not conducive to the continuous growth of the wax crystal to form a network structure, so it has excellent anti-wax effect [2-4]. In this paper, a series of paraffin inhibitors of polyacrylic acid high-carbon alcohol ester were synthesized. Polyhexadecyl acrylate was selected as paraffin inhibitor by static paraffin inhibition method, and the synthesis conditions, dosage and other factors were investigated.

2. Experiment part

2.1. Synthesis of high-carbon acrylate

High-carbon alcohols (dodecanol, tetradecanol, hexadecanol and octadecanol) were added to the three-port flask and completely melted by heating. Then 1.2 times the amount of acrylic acid, 0.6 % inhibitor and 1 % catalyst were added and heated to 110 °C for 6 hours. After the reaction, cool, add ethyl acetate, wash with saturated sodium carbonate solution to weak alkaline, and then wash with saturated sodium chloride solution to neutral, separate the organic layer, add anhydrous calcium chloride to dry, filter, evaporate the solvent to obtain acrylic high carbon alcohol ester.

2.2. Synthesis of polyacrylic acid higher alcohol ester

Acrylate high-carbon alcohol ester was added to a three-necked flask (with or without a certain amount of crosslinking agent N,N'-dimethyleneacrylamide), dissolved in toluene, and then added with 0.1 % initiator. The reaction was refluxed under stirring, cooled to room temperature, and the solvent was evaporated. Poly (dodecanol acrylate) (PAD), poly (tetradecyl acrylate) (PAT), poly (hexadecyl acrylate) (PAH), poly (octadecyl acrylate) (PAO) and crosslinked poly (dodecanol acrylate) (CPAD), crosslinked poly (tetradecyl acrylate) (CPAT), crosslinked poly (hexadecyl acrylate) (CPAH), crosslinked poly (octadecyl acrylate) (CPAO) were obtained.

2.3. Determination experiment of wax prevention rate

Accurately weigh 100 g paraffin and 300 g 0# diesel in a 500 ml beaker, put it in an electric thermostatic water bath, control it at 50 °C, and stir it until the paraffin is completely dissolved and sealed for later use. Select multiple iron sheets with similar external surface area, smooth the surface until the surface is rust-free, and then put them in alcohol. After a few hours, it was taken out and dried. Three pieces with similar surface area were selected and numbered, weighed and recorded respectively. Method for determination of wax prevention rate (static wax prevention method) [6]: Accurately weigh the quality of steel sheets (including fixed fine lines), measure 50 ml of wax-containing oil in an iodometric bottle, add a certain amount of wax prevention agent, steel sheets are suspended in oil, sealed with a lid, at a constant temperature of 60 °C for 1 h, naturally cool to 30 °C, remove the steel sheet and accurately weigh the quality of the steel sheet. The wax inhibition rate η is calculated by the following formula.

$$\eta = (m_1 - m_2)/(m_1 - m_0) \times 100\%. \quad (1)$$

where η is the wax prevention rate, %; m_0 is the mass of steel sheet, m_1 is the mass of steel sheet without wax inhibitor, and m_2 is the mass of steel sheet with wax inhibitor.

3. Results and discussion

3.1. Synthesis and optimisation of polyacrylic acid polycarbol ester anti-waxing agents

The carbon chain length of high-carbon alcohols in monomers is the basic factor affecting the antifouling effect. Firstly, the antifouling agents with different monomer polymerisation were selected, and their dosage was fixed at 150 mg/L to evaluate and compare the antifouling effect. The results are shown in Fig.1. From the figure, it can be seen that the antifouling effect of poly(cetyl acrylate) was the best, 84.5%. Therefore, the synthesis conditions of poly(hexadecyl acrylate) were optimized.

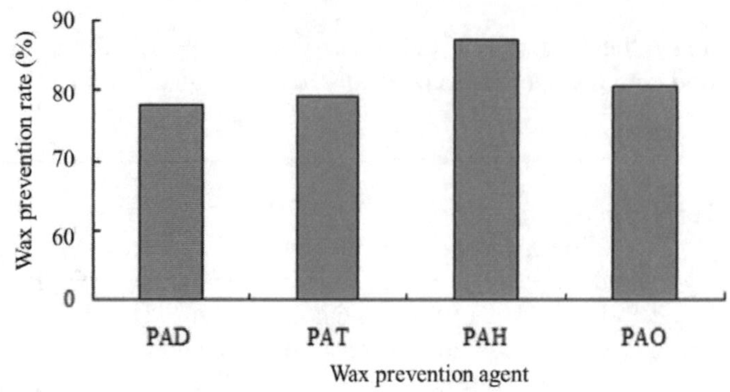

Figure 1. Anti-waxing effect of anti-waxing agent.

3.1.1. Effect of reaction temperature on the anti wax effect of polyacrylic acid hexadecanol ester. In the free radical polymerization reaction of olefinic monomers, the reaction temperature mainly affects the decomposition rate of the initiator. The lower the reaction temperature, the slower the decomposition rate of the initiator, and the larger the molecular weight of the generated polymer. There is a certain relationship between the molecular weight of the polymer and its anti wax effect. Therefore, the anti wax rate of polyacrylic acid hexadecanol ester synthesized at different reaction temperatures (reaction time 6 hours) was experimentally studied, and the results are shown in Figure 2. As can be seen from the figure, when the reaction temperature is 60 °C, the wax inhibitor obtained has the highest wax prevention rate. Therefore, in the following synthesis, the reaction temperature is set at 60 °C.

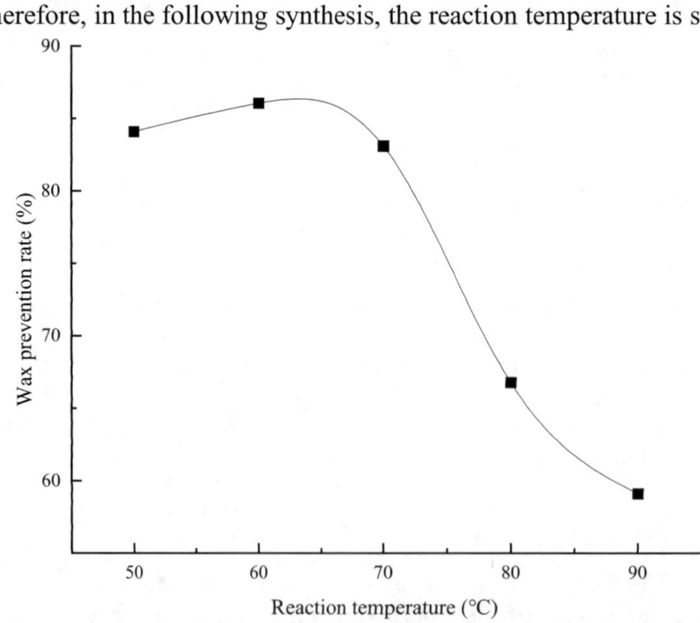

Figure 2. The effect of reaction temperature on wax prevention effect.

3.1.2. Effect of reaction time on the anti wax effect of polyacrylic acid hexadecanol ester. The reaction time affects the degree of polymerization and conversion rate, and the effect of reaction time on the wax prevention rate of polyacrylic acid hexadecanol ester is shown in Figure 3. As shown in the figure, after the reaction time reaches 4 hours, the anti wax rate of the polymer basically does not change and is above 86%. Therefore, the reaction time is selected as 4 h.

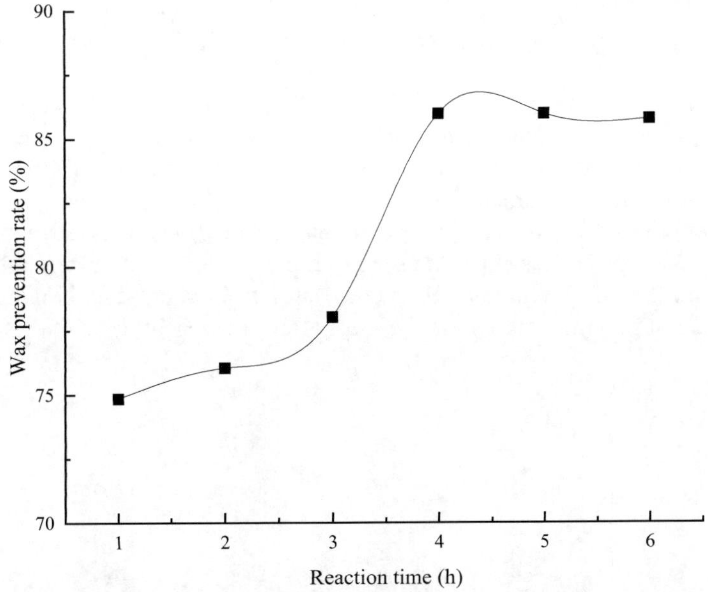

Figure 3. The effect of reaction time on wax prevention effect.

3.1.3. The effect of polyacrylic acid hexadecanol ester dosage on the anti wax effect. The amount of anti wax agent has a significant impact on the anti wax rate, and the anti wax effect of polyacrylic acid hexadecanol ester at different amounts is shown in Fig.4. As shown in the figure, as the amount of wax inhibitor increases, the wax prevention rate also increases accordingly. When the amount of wax inhibitor is 150 mg/L, the wax prevention rate reaches 89.3%. Continuing to increase the dosage, the wax prevention rate remains basically unchanged, so the optimal dosage is 150 mg/L. The low wax prevention rate at low dosage is due to the insufficient amount of action to effectively disperse wax crystals. Before the eutectic effect is exerted, a large amount of wax crystals have already precipitated, and adsorption can only be used to disperse wax crystals, which inevitably has a limited effect, thus affecting the wax prevention effect [5].

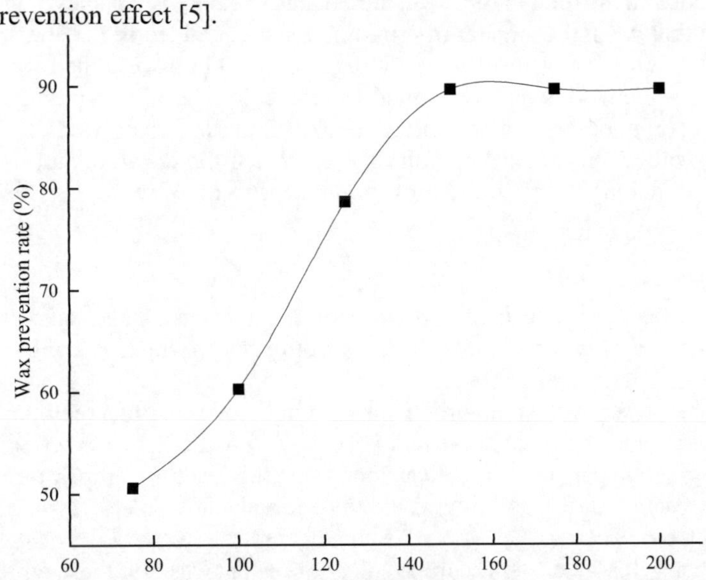

Figure 4. The effect of the amount of anti-wax agent on the anti-wax rate.

3.2. Anti-waxing effect of N,N'-dimethylene acrylamide-crosslinked poly(acrylic acid) polycarbol esters

N,N'-dimethylene acrylamide was applied as a cross-linking agent to synthesize polymers with a reticulated structure, which is conducive to preventing wax deposition by preventing paraffin wax intermolecular connections from forming gels or agglomerating into clumps. Therefore, we investigated the wax prevention effect of N,N'-dimethylene acrylamide crosslinked polyacrylic acid polycarbonate esters. The wax prevention effect of poly(high carbonyl acrylate) synthesized with 1% cross-linking agent is shown in Fig. 5, from which it can be seen that the wax prevention rate increased to different degrees compared with that of poly(high carbonyl acrylate) with no cross-linking agent, and the cross-linking poly(hexadecanol poly(acrylate)) was the most effective, reaching 82.4%.

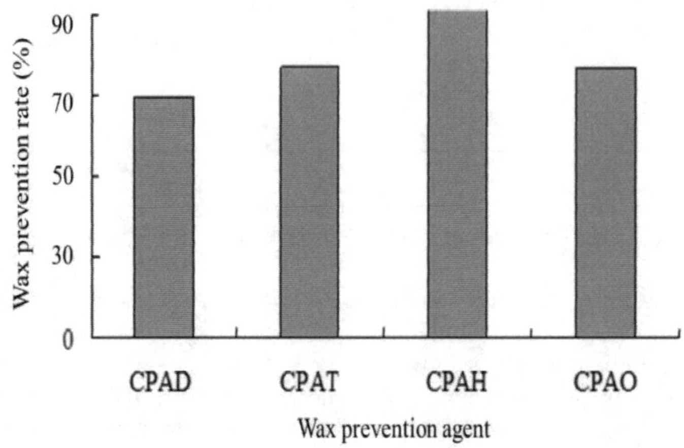

Figure 5. The anti wax effect of different cross-linked polyacrylic acid high carbon alcohol ester.

The amount of crosslinking agent determines the mesh size of the network structure, which also affects the deposition effect of wax. Therefore, the amount of crosslinking agent (the percentage of crosslinking agent added to the total mass) was studied, and the results are shown in Figure 6. From the figure, it can be seen that when the amount of crosslinking agent added is 2%, the anti wax effect is the best, reaching 98.4%; Excessive or insufficient dosage can lead to a weakened wax prevention effect, which may be due to the network structure formed when the dosage of crosslinking agent is 2%, which is suitable for the precipitation of paraffin on the network. Its long chain leads to loose and dendritic crystal pile masonry, preventing the deposition of wax. When the dosage is too large, it may lead to dense crystal stacking, making it difficult to prevent the deposition of wax.

4. Conclusion

(1) We synthesized polyacrylic acid high carbon alcohol ester and characterized the structure of the synthesized monomer and polymer using infrared spectroscopy, which is consistent with the target compound.
(2) The effects of factors such as the carbon chain length of monomer compounds, the synthesis conditions of polymers, and the amount of anti-wax agent on the anti-wax effect were investigated through static wax formation experiments. It was found that the reaction temperature of poly (hexadecyl ester) was 60 °C, the reaction time was 4 hours, and the optimal anti wax effect was achieved at a dosage of 150 mg/L, reaching 89.3%.
(3) A network polyacrylic acid high carbon alcohol ester was synthesized by adding N, N'-dimethylacrylamide as a crosslinking agent. The cross-linked polyacrylic acid hexadecanol ester was selected as an anti-wax agent, and the amount of crosslinking agent was investigated. The best anti wax effect was achieved at a dosage of 2%, reaching 98.4%.

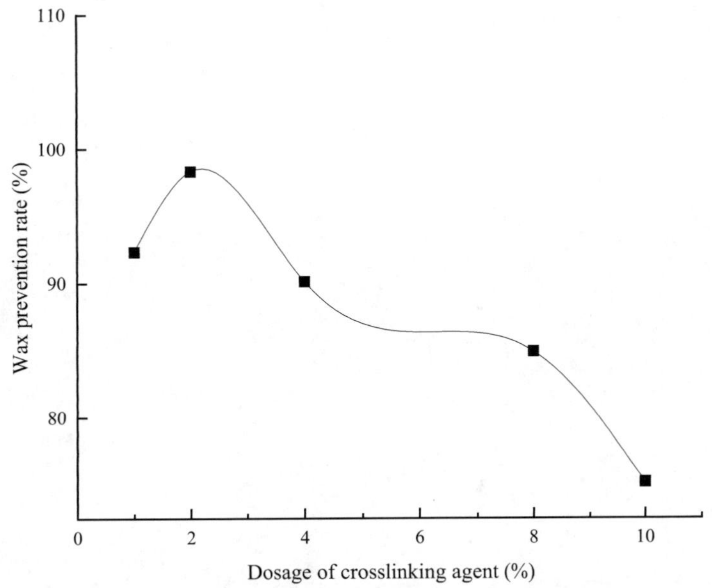

Figure 6. The influence of the dosage of crosslinking agent on the wax prevention effect.

Acknowledgements
This work was financially supported by grants from Youth Innovation Team of Shaanxi University, Shaanxi Key Research and Development Plan (2023-YBGY-052) and Scientific Research Program Funded by Shaanxi Povincial Education Department (22JY052).

References

[1] Dajun C and Fu C. 2006 *Applied Chemistry in Oil and Gas Fields.* (Beijing: Petroleum Industry Press) **6** 223-248.
[2] Yu Z. 2006 Development of YS-3 paraffin remover and anti wax agent. *Journal of Petroleum and Natural Gas.* **28** (9) 136-137.
[3] Zhichun W, Xiuli Z and Yuhua Z 2001 Development and application of polymer flooding viscosity reducing and wax preventing agent. *Oil and Gas Field Surface Engineering (OGSE).* **20** (3) 38-39.
[4] Gang L, Zhong Y and Dajun C 2005 Study on the wax inhibition rate of star shaped polymer wax inhibitors with different structures. *Applied Chemistry.* **34**(1) 45-48.
[5] Juntao M and Zhiyu H. 2001 The development of polymer anti wax agents and the effect of their structure on their performance. *Journal of Xi'an Petroleum Institute (Natural Science Edition).* **16**(4) 45-48.
[6] Jiuming L, Xiaoling J and Feng C 2002 Development of CRT-2 polymer wax remover. *Petroleum and Natural Gas Chemical.* **30**(4) 202-210.
[7] Bo Y, Haoran Z and Baodong Y 2023 Pre-heating temperature induced flowability and wax deposition characteristics of crude oil adding wax inhibitors. *Petroleum Science.* **20**(4) 2468-2478.
[8] Zhao Z, Liu D and Ren X 2023 Effect of EVA wax inhibitor on the W/O emulsion stabilized by asphaltenes and paraffin wax. *Colloids and Surfaces A: Physicochemical and Engineering Aspects.* **659**(2) 130642.

AUTHOR INDEX

Institute of Physics
Dirac House, Temple Back
Bristol BS1 6BE UK

ISSN: 1742-6588
ISBN 978-1-7138-9278-6

International Scientific and Practical Conference “Sustainable Development of Traditional and Organic Agriculture in the Concept of Green Economy” (SDGE 2021)

BIO Web of Conferences Volume 42 (2022)

Smolensk, Russia
27 November 2021

edp sciences